# e-food

## Food and technology

### Book 1

**Leanne Compton**  **Carol Warren**

OXFORD

# OXFORD
UNIVERSITY PRESS

253 Normanby Road, South Melbourne, Victoria 3205, Australia

Oxford University Press is a department of the University of Oxford.
It furthers the University's objective of excellence in research, scholarship,
and education by publishing worldwide in

Oxford  New York

Auckland  Bangkok  Buenos Aires  Cape Town  Chennai
Dar es Salaam  Delhi  Hong Kong  Istanbul  Karachi  Kolkata
Kuala Lumpur  Madrid  Melbourne  Mexico City  Mumbai  Nairobi
São Paulo  Shanghai  Taipei  Tokyo  Toronto

OXFORD is a trade mark of Oxford University Press
in the UK and in certain other countries

© Leanne Compton, Carol Warren 2002

First published 2002
Reprinted 2003, 2004

This book is copyright. Apart from any fair dealing for the purposes
of private study, research, criticism or review as permitted under the
Copyright Act, no part may be reproduced, stored in a retrieval system,
or transmitted, in any form or by any means, electronic, mechanical,
photocopying, recording or otherwise without prior written permission.
Enquiries to be made to Oxford University Press.

**Copying for educational purposes**

Where copies of part or the whole of the book are made under Part VB
of the Copyright Act, the law requires that prescribed procedures be
followed. For information, contact the Copyright Agency Limited.

National Library of Australia
Cataloguing-in-Publication data:

Compton, Leanne
  e-food.

  Includes index.
  For secondary students.
  ISBN 0 19 551485 8.

  1. Food—juvenile literature. 2. Nutrition—juvenile
  literature. 3. Quick and easy cookery—juvenile
  literature. I. Warren, Carol Joan. II. Title.

641.3

Typeset by Sylvia Witte
Printed by Bookpac Production Services, Singapore

**OWLS**
OXFORD DICTIONARY WORD AND LANGUAGE SERVICE

Do you have a query about words, their origin, meaning, use, spelling, pronunciation, or any other aspect of international English? Then write to OWLS at the Australian National Dictionary Centre, Australian National University, Canberra ACT 0200 (email ANDC@anu.edu.au). All queries will be answered using the full resources of *The Australian National Dictionary* and *The Oxford English Dictionary*.

The Australian National Dictionary Centre and Oxford University Press also produce OZWORDS, a biannual newsletter which contains interesting items about Australian words and language. Subscription is free – please contact the OZWORDS subscription manager at Oxford University Press, GPO Box 2784Y, Melbourne, VIC 3001, or ozwords@oup.com.au

# CONTENTS

Introduction  *iv*
The Technology Design Process  *v*
Acknowledgments  *vi*

## Section 1  READY, SET, GO!  *1*
Chapter 1  Get Going: Discover Food and Technology  *2*
Chapter 2  Decisions! Decisions!: Deciding What to Eat  *12*
Chapter 3  Supermodels: Discover Healthy Eating Models  *23*

## Section 2  CEREALS, BREAD, RICE, PASTA AND NOODLES  *31*
Chapter 4  Let's Get Cereals: Discover Grains  *32*
Chapter 5  D'oh!: Discover Bread  *41*
Chapter 6  On the Boil: Discover Rice  *50*
Chapter 7  Fiddlesticks, Chopsticks: Discover Pasta and Noodles  *64*

## Section 3  VEGETABLES AND LEGUMES  *73*
Chapter 8  Veg Out: Discover Vegetables  *74*
Chapter 9  Bean There, Done That!: Discover Legumes  *89*

## Section 4  FRUIT  *101*
Chapter 10  Frrrruit!: Discover Fruit  *102*

## Section 5  MILK, YOGHURT AND CHEESE  *111*
Chapter 11  Milk It!: Discover Milk  *112*
Chapter 12  Need a Little Culture?: Discover Yoghurt  *120*
Chapter 13  Say Cheese: Discover Cheese  *129*

## Section 6  MEAT, FISH, POULTRY, EGGS, NUTS AND LEGUMES  *137*
Chapter 14  Steak Your Claim: Discover Meat  *138*
Chapter 15  Sounds Fishy: Discover Seafood  *148*
Chapter 16  One for the Birds: Discover Poultry  *156*
Chapter 17  Egged On: Discover Eggs  *166*
Chapter 18  Going Nuts: Discover Nuts  *174*
Chapter 19  Pulses Rising: Discover Vegetarians  *182*

## Section 7  WATER  *189*
Chapter 20  Water Works!: Discover Water  *190*

## Section 8  EAT IN SMALL AMOUNTS  *197*
Chapter 21  Fat Is Not a Four-letter Word: Discover Fats  *198*
Chapter 22  Sweet!: Discover Sugar  *201*
Chapter 23  Salt of the Earth: Discover Salt  *208*

Glossary  *215*
Index  *216*
Recipe Index  *218*
CSF II Grid

# INTRODUCTION

*e-Food Book 1* aims to encourage a fun, interactive and practical approach to the study of food through Home Economics for junior secondary school students. The title *e-Food* reflects the impact of technology and the electronic age on food.

The 'Ready, Set, Go!' section takes an activity-based approach to the teaching of safety, hygiene, food models and choices. The food groups from *The Australian Guide to Healthy Eating* have been used to structure the remaining chapters, which provide interesting information about food from a nutritional, historical and contemporary perspective.

The activity-based approach is reflected throughout the book as you will find a broad range of activities, such as revision questions, investigations, puzzles, newspaper articles and case studies that particularly encourage the use of the Internet and multimedia. The inclusion of assessment tasks, addressing outcomes from the Technology and Health and Physical Education Key Learning Areas, is also a unique feature of the book. A CD-ROM is also available and includes black-line masters with answers for many of the activities throughout the book, assessment criteria sheets and additional activities.

Recipes, which form a major component of the book, have been developed to reflect our multicultural, busy lifestyles while promoting that food should be easy to prepare, fresh and nutritionally good for us.

The broad range of activities and visual appeal throughout *e-Food Book 1* has been designed to offer plenty of choice for teachers to engage students with a range of abilities and learning styles. We hope that you will find this book to be an exciting educational resource that will help to meet the needs of our young people in a technologically diverse and challenging world, and that students will have as much fun using this book as we did writing it.

# THE TECHNOLOGY DESIGN PROCESS

The Technology Design Process involves the four phases of investigating, designing, producing and evaluating.

When you research or gather information, you are investigating. This can be done by searching on the Internet, reading books or listening to your teacher or other people.

Designing involves considering your options and making decisions about how you will go about your tasks, which resources you will use and how you will utilise your creative ideas to meet the requirements of the task (design brief).

Producing is when you undertake the task or make a product. This could be in the form of practical work, when you complete some of the recipes in this book or when you complete a project or assignment.

Evaluating is when you look back and reflect on the decisions made, the outcomes and how you might do things differently next time.

All phases involve making decisions, managing situations or solving problems. There is not necessarily a particular sequence to the phases of the Technology Design Process; however the four phases often flow sequentially, and it is more likely that they overlap and repeat continually. This is seen in the diagram below, which shows a continuous, flowing yet random connection between the four phases.

> **e-fact**
> A design brief is a defined situation, usually in the form of a statement, that outlines guidelines for a task.

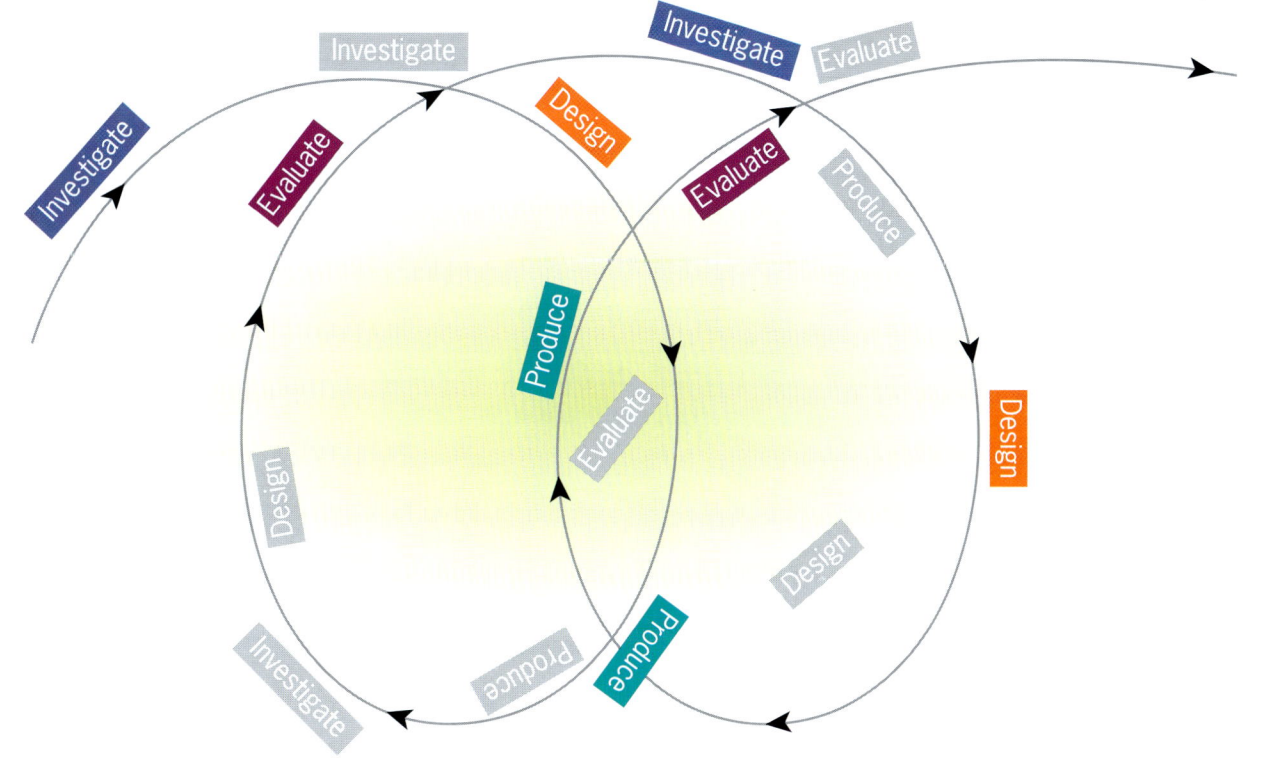

# Acknowledgments

The authors and publisher wish to thank the following copyright holders for granting permission to reproduce their material:

AAP: p. 136. Ananova: p. 170 © Ananova Ltd. 2002. Reproduced by permission. All rights reserved. Australia New Zealand Food Authority: p. 17; Australian Pork Corporation: p. 142. Carolyn Holbrook: pp. 81–82. *Daily Mail*: p. 11. Dick Smith Foods: p. 104. Fairfax Photo Library: p. 116. Family Planning Victoria: p. 80. The *Herald & Weekly Times* Photographic Collection: pp. 7, 9, 10, 19, 143. Info Access: pp. 27–28 © Commonwealth of Australia. Reproduced by permission. Meat & Livestock Australia: pp. 140–141. National Health and Medical Research Council (NHHRC): pp. 25–26. Nutrition Australia: p. 24 (top, bottom left). Oldways Preservation & Exchange Trust: p. 24 (bottom right). photolibrary.com: pp. 67, 166, 167, 202. Pauls Limited: p. 122 (left). Rita's Nut Shop, South Melbourne Markets: p. 89. Sanitarium: p. 176.

Every effort has been made to trace the original source of copyright material contained in this book. The publisher would be pleased to hear from copyright holders to rectify any errors or omissions.

## Section 1

# Ready, Set, Go!

# Chapter 1

# Get Going: Discover Food and Technology

Here are a series of activities to begin your exploration not only of food and the Food Technology Centre but also of issues concerning safety, hygiene, sensory evaluation and nutrition.

## ▶ Safety sleuth

Investigate the safety concerns in the Food Technology Centre or your kitchen at home. Write up a report, suggesting how these safety considerations could be improved or overcome. Think about how to best present your work. For example, you might like to use a word processing or design program with graphics.

GET GOING

## ▶ Food and technology bingo

The objective of 'Food and technology bingo' is to get a row of squares completely signed. The row can be completed diagonally, vertically or horizontally.

A different classmate must sign each square. Who can get a row of squares signed first? Who can get the most squares signed?

Good luck!

# BINGO

| Loves to eat chocolate | Reads food labels | Favourite food is ice cream | Is a vegetarian | Knows someone who is a coeliac (allergic to gluten/wheat) |
|---|---|---|---|---|
| Cooked dinner in the last week | Wants to work in the hospitality industry | Has toast for breakfast | Does not like McDonald's | Loves Thai food |
| Eats fruits every day | Knows which vitamin is found in oranges<br><br>Answer:_____ | Drinks water every day | Does not put salt on their food | Loves to prepare food |
| Drinks soy milk or soy products | Spreads margarine or butter thinly on their bread/toast | Eats meat every day | Eats pasta regularly | Helps their parents with the shopping |
| Knows the most important meal of the day<br><br>Answer:_____ | Loves eating | Consumes breakfast cereal as a snack | Regularly eats vegetables in their diet | Knows the main mineral found in milk<br><br>Answer:_____ |

READY, SET, GO!

## ▶ Mix and match processes

Match each of the processes in the first column of the table below with their correct definition in the second column.

| Process | Definition |
|---|---|
| Beat | Shake dry ingredients through a sieve to remove the lumps and aerate. |
| Blend | Combine ingredients evenly. |
| Boil | Decorate food. |
| Chop | Manipulate dough by folding and turning so that it becomes smooth. |
| Cream | Beat liquid until slightly foamy. |
| Dice | Beat sugar and butter until the sugar dissolves. This makes the mixture look like whipped cream: it is light and fluffy and pale in colour. |
| Garnish | Brush liquid on the surface of food to create a glossy brown appearance. |
| Glaze | Blend food with a processor until smooth. |
| Grill | Cut into small cubes. |
| Julienne | Mix dry ingredients with a moist ingredient to form a smooth paste. |
| Knead | Place food in a liquid to absorb the flavours. |
| Marinate | Cut food into fine strips, like matchsticks. |
| Mix | Cut food thinly. |
| Purée | Heat liquid until bubbles rise rapidly to the surface. |
| Rub in | Mix butter or margarine into flour with fingertips until mixture looks like breadcrumbs. |
| Sift | Cook food in oil or butter over a low heat. |
| Simmer | Mix rapidly in a circular manner. |
| Slice | Cook food using dry radiant heat, such as under a griller. |
| Sauté | Bring liquid to just below boiling point—bubbles will appear on the surface. |
| Whisk | Cut food roughly. |

Can you think of other processes used in recipes? Add them to your glossary. Find recipes that contain each of the processes mentioned above.

## ▶ A–Z equipment brainstorm

Identify a piece of equipment in the Food Technology Centre that begins with each letter of the alphabet.

## ▶ Equipment treasure hunt

The table below includes a sketch of a measuring cup and explains its use. Draw up a similar table in your workbook. Sketch and explain the use of each of the following pieces of equipment: measuring jug, rolling pin, fork, skewer and sifter. Include any other pieces of equipment you can think of in your table.

| Equipment name | Sketch of equipment | Uses |
|---|---|---|
| Measuring cup ($\frac{1}{2}$ cup measure) | | To provide an accurate level measure of dry ingredients. |

GET GOING

## ▶ Stoves scrutiny

Research the types of stoves in the Food Technology Centre. Draw a floor plan of the centre and identify which stoves are gas or electric. Draw one stove and label the different parts. Write down a list of instructions, outlining how to turn the stove on safely.

## ▶ Abbreviation mix and match

Match each of the abbreviations in the first column of this table with the correct word in the second. Read through several recipes and identify other abbreviations. What do they stand for?

| Abbreviation | Word |
|---|---|
| tb | Centimetre |
| tsp | Litre |
| c | Self-raising |
| kg | Degrees Celsius |
| g | Gram |
| mL | Kilogram |
| min | Packet |
| SR | Teaspoon |
| L | Tablespoon |
| °C | Cup |
| cm | Minute |
| pkt | Millilitre |

## ▶ Bacteria growth

Fill Petri dishes with about 1.5 per cent nutrient agar and collect samples from sinks, floors, tea towels and benchtops, as well as breathing, touching and placing samples of hair on some of the dishes. Report on the growth of bacteria that appears within a few days. Write up a report about your findings. What conclusions can you draw?

## ▶ Small equipment exercise

In small groups, investigate one of these small appliances:
- food processor
- juicer
- handheld electric beater
- microwave oven
- electric frypan
- blender
- toaster
- kettle

In your investigation, include the following information:
- pictures/diagrams of appliance
- its uses
- its cost
- recipes that use the appliance
- safety considerations

READY, SET, GO!

## ▶ Recipes

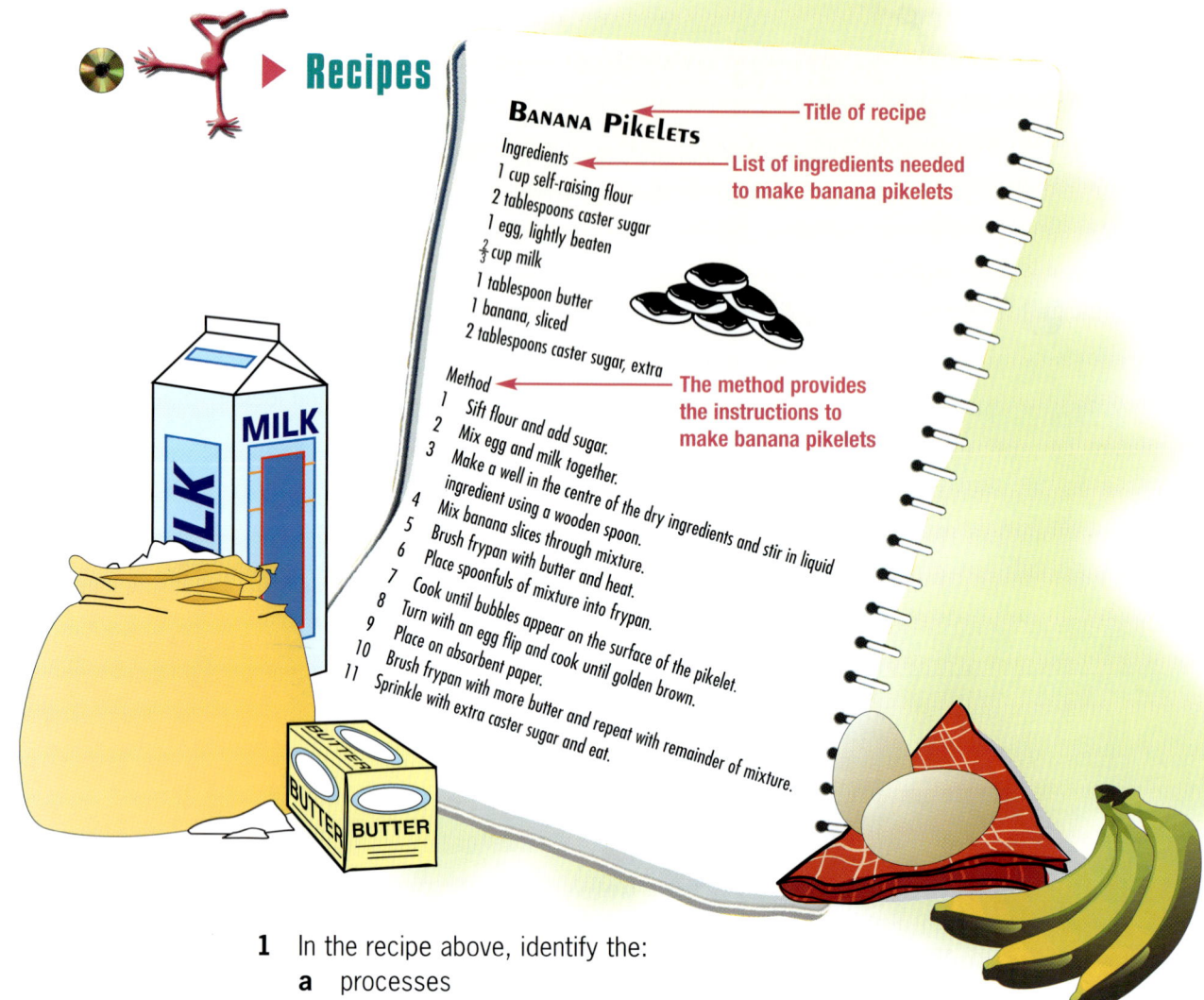

### Banana Pikelets

*Title of recipe*

**Ingredients** ← List of ingredients needed to make banana pikelets

1 cup self-raising flour
2 tablespoons caster sugar
1 egg, lightly beaten
2/3 cup milk
1 tablespoon butter
1 banana, sliced
2 tablespoons caster sugar, extra

**Method** ← The method provides the instructions to make banana pikelets

1. Sift flour and add sugar.
2. Mix egg and milk together.
3. Make a well in the centre of the dry ingredients and stir in liquid ingredient using a wooden spoon.
4. Mix banana slices through mixture.
5. Brush frypan with butter and heat.
6. Place spoonfuls of mixture into frypan.
7. Cook until bubbles appear on the surface of the pikelet.
8. Turn with an egg flip and cook until golden brown.
9. Place on absorbent paper.
10. Brush frypan with more butter and repeat with remainder of mixture.
11. Sprinkle with extra caster sugar and eat.

**1** In the recipe above, identify the:
  **a** processes
  **b** abbreviations
  **c** dry ingredients
  **d** wet ingredients
**2** Why do you think it is important to carefully read the recipe prior to preparation?

## ▶ ⓔ-Safety links

Visit each of the websites below for information on food hygiene and safety. Then complete the associated tasks.

**www.foodscience.afisc.csiro.au/consumer.htm**

Research one of these topics:
- safety of microwave ovens
- handling of food in the home
- storage life of foods

**www.foodsafety.vic.gov.au**
- Here you can find further information about Victorian food laws. Write a newspaper article outlining the food laws. Think about how to best present

GET GOING

your work. You might like to use a word processing or design program with graphics.
- Construct a table to outline factors that assist with keeping food safe.

### www.safefood.net.au

Here you will find lots of great information to explore food hygiene and safety. There are links to:
- fact sheets
- food safety tips
- fun and games, such as Hangman

### www.woolworths.com.au/dietinfo/rsa27.asp

- Write a report, outlining the safety and hygiene considerations when using such foods as:
    - meat, fish and chicken
    - fruit and vegetables
    - eggs
    - dairy foods
    - food from delicatessens
    - leftovers
- Create a pamphlet to provide information on how food can be safely brought home and stored correctly. Think about how best to present your work. You might like to use a design program with graphics.
- Draw a poster to illustrate how you should clean up safely.

## ▶ Kitchens come in from the cold

Read the newspaper article and then answer the questions that follow it.

---

# Kitchens come in from the cold

**KITCHENS KARMA**

How we've changed

**IN**
- Stainless steel, commercial look
- Slide-out pantries
- Microwaves
- Dishwashers
- Built-in coffee machines
- Breadmakers
- Pullout rangehoods
- Ice-making machines
- TV & stereo speakers

**OUT**
- Walk-in pantries
- Waste-disposal units & rubbish compactors
- Large rangehoods
- Hot water dispensers/urns
- Visible rubbish bins
- Double sinks
- Freestanding stoves
- Lead-lighting

DB/Herald Sun 17/8/2001

**BY JON RALPH**

MICROWAVES, dishwashers and coffee machines have all had an impact on Australian kitchens in the past 30 years.

But despite homeowners' obsession with time-saving gadgets, experts say it is the structure and role of Australian kitchens that have changed considerably.

Direct Kitchen director John Sullivan said modern kitchens were all about intergrating cooking with entertaining or relaxing.

'At one stage the kitchen was a facility for preparing food,' he said. 'Now the room is a space for cooking, handling entertainment, looking good and involving everyone in the cooking process.'

He said the microwave and dishwasher had clearly made the most inroads in the past 30 years.

'Microwaves have been the big issue. They have been a recent development but almost everyone would have one. I would say 80 per cent of households have one.'

Australian kitchens are now firmly entrenched as part of the entertainment area, according to Max Pesch, marketing manager of cooking equipment company Ilve.

'Kitchens are designed to be part of the leisure entertainment, whereas they used to be a separate room out the back,' he said.

The types of food we cook also had an influence on our equipment.

The kitchen's role will be featured at Melbourne's Home Show and Garden Expo. The show celebrates 30 years this year by taking a look into the past, with the theme Relivin' the 70s.

Running from tomorrow until August 26, the show will also glance into the future, with Melbourne architects and interior designers revealing what they think homes and gardens will look like in 2031.

HOME Expo info line: 1900 931 798.

READY, SET, GO!

1. According to the article, in what ways have the structure and role of the kitchen changed?
2. What types of equipment are 'in'?
3. What types of equipment are 'out'?
4. Explain how the types of food we cook influence which equipment we use? Refer to specific examples.

## ▶ Sensory evaluations

Appearance, flavour, texture, aroma

- Appearance: How food looks.
- Flavour: What food tastes like.
- Texture: How food feels in your mouth.
- Aroma: How food smells.

In the table below, numerous terms are listed for appearance, flavour, texture and aroma (AFTA).

| Appearance | Flavour | Texture | Aroma |
| --- | --- | --- | --- |
| Colourful | Bitter | Fatty | Fruity |
| Golden | Bland | Hard | Sweet |
| Well shaped | Burnt | Crisp | Spicy |
| Well risen | Creamy | Doughy | Yeasty |
| Uneven | Tangy | Sticky | Mild |
| Undercooked | Sweet | Lumpy | Strong |
| Glossy | Sour | Grainy | Mild |
| Dull | Strong | Smooth | Sour |
| Cloudy | Hot | Grainy | Rancid |
| Transparent | Spicy | Creamy | Garlicky |
| Curdled | Salty | Crunchy | |
| Glazed | Oily | Soft | |
| Dull | Tart | Coarse | |

1. Can you add five more terms to each of the categories?
2. Describe the AFTA of these foods:
    a  dry biscuit
    b  chocolate ice cream
    c  apple
    d  nuts
    e  sausage roll
    f  salami
    g  strawberry jelly
    h  tofu
    i  potato crisps
    j  sweet chilli sauce

## ▶ Trick or two helps the food go down

Read the newspaper article and then answer the questions that follow it.

## Trick or two helps the food go down

THERE is a correct way to eat every tricky foodstuff, according to the Elly Lukas Beauty Therapy Modelling and Deportment School.

Peas should be pushed on to the downturned fork by the knife then transferred carefully to the mouth. Scooping them with an upturned fork is not acceptable, says teacher Janelle Sutton.

Artichokes can only be eaten using the fingers. However, a finger bowl should be provided and used by dipping the fingers only, then drying them discreetly on the napkin.

Corn on the cob should be anchored by special cob-holders and gnawed. However, cobs are really only appropriate for informal parties or family dinners.

Halved avocados that have been filled with vinaigrette or seafood are eaten with a small spoon provided with the dish. It is permissible to hold the avocado with one hand to prevent it slipping around.

Bread rolls should be broken by inserting the thumb, tearing off small pieces with the fingers and buttering each piece as it is eaten. Do not butter the entire roll.

Whole fish is eaten by removing the skin and setting to one side. Eat the top flesh, then use the knife to nip the backbone at the fish's tail and lift it out with the knife and fork. Set it at the side of the plate and eat the rest of the flesh.

Pasta, despite being messy, is worth the effort. Long noodles, such as fettuccine or spaghetti, are dealt with by separating out a few strands and winding them around the fork by pressing it against the side of the spoon.

**1** Describe three rules for eating food that are identified in the article.

**2** List five other rules that your family has in regard to eating meals.

### ▶ Study spotlights poor health

Read the newspaper article and then answer the questions that follow it.

## Study spotlights poor health

**By WENDY BUSFIELD, medical reporter**

AUSTRALIA faces an epidemic of diabetes, obesity and heart disease.

Medical tests on more than 6000 adults in four states show Australia has one of the highest diabetes rates in the world.

One in every four adults either suffers from diabetes or has a high risk of developing the debilitating disease, according to the national study.

The average Australian has also gained 5 kg in the past two decades, with 60 per cent of adults either obese or overweight.

Research leaders Paul Zimmett and Tim Welborn said Australia had jumped to the same ranks as the United States.

They said the study results were a grave warning of the massive human and financial cost of poor health.

Professor Zimmett said the number of people with diabetes had doubled in the past two decades, with worse figures to come.

He said Australia had a sporting reputation, but the results showed more work was needed on health promotion at all levels of the community.

Prof. Welborn was particularly concerned by the 15 per cent rise in obesity rates.

'In 20 years we have become a fat nation,' Prof. Welborn said.

'The results left no room for complacency and placed diabetes as one of the highest public health priority issues for the nation, particularly because of the resulting burden of heart complications, stroke, blindness, kidney failure and amputations.'

The Australian Diabetes, Obesity and Lifestyle Study gives health authorities the first national snapshot of risk factors for diabetes and heart disease.

The results show seven out of 10 Australians over 25 have at least one out of the four risk factors—glucose intolerance, obesity, high blood fats or hypertension.

The study—which included surveys in Victoria, Western Australia, Tasmania and New South Wales—also revealed:

**MORE** than 7 per cent of over 25-year-olds have diabetes.

**ANOTHER** 16 per cent in the age group are at high risk of diabetes, with impaired glucose metabolism.

**ONE** in five Australians over 65 has diabetes.

'For every known case of diabetes there is one person with the disease who doesn't know they have it,' the study concluded.

**1 a** List the three dietary-related diseases outlined in the article.
  **b** List two foods that may contribute to each disease.

**2 a** From the article, how many kilograms has the average Australian gained in the past twenty years?
  **b** What factors might have contributed to this trend?

**3 a** What percentage of the adult population is overweight or obese?
  **b** What is the difference between the terms *overweight* and *obese*?

READY, SET, GO!

    **c** In the article Professor Welborn says that Australia is a 'fat nation'. Do you agree? Explain your answer.

    **d** Give three strategies to assist with maintaining a healthy weight.

**4** List the health problems outlined that may result from diabetes.

**5** How could studying this book assist you with promoting good health?

## ▶ Pea-sized problem

Read the newspaper article and then answer the questions that follow it.

---

# Pea-sized problem

### A vegie rolls to oblivion
**By JULIA COFFEY**

*I eat my peas with honey,
I've done it all my life.
It makes my peas taste funny,
But it keeps them on my knife.*
—SPIKE MILLIGAN

HONEY may be funny, but the art of getting peas on to the fork and into the mouth is no laughing matter.

According to the experts, today's children are so inept with cutlery that they no longer eat peas.

Scientists have been asked to develop a bigger, child-friendly pea.

The problem is that many children are growing up without being taught table manners.

Fast food and busy lifestyles mean many families no longer sit down at the table for dinner. Instead, children are eating their meals using their fingers.

So when boffins at a British supermarket discovered most of its fresh peas were being consumed by adults, they decided to look further.

They found the growing use of short-cut food by time-starved parents meant children were becoming used to eating without knife and fork.

The supermarket chain has asked scientists to develop a large easy-to-eat pea to encourage parents to reintroduce the vegetable to their dinner plates.

A recent report by the Bureau of Statistics found households spent an average of $18.80 of their weekly income on fast food and takeaways—one of the biggest single expenditures.

Nuts and fruit, both fresh and dried, accounted for only $7.82, while households spent just $8.72 on vegetables each week.

Karen Inge, dietitian with the Institute of Health and Fitness, said the dying art of table manners was symptomatic of a shift in eating habits.

Busy lifestyles meant families ate on the run, microwaves allowed meals to be taken at different times and children often ate in front of the television.

Ms Inge said the loss of table manners could become an obstacle to success in adult life.

'If you go to a restaurant to sit down and eat in a business meeting or with guests, you should know what's appropriate and how to eat appropriately in certain circumstances,' she said.

Table manners are a part of the curriculum at the Elly Lukas Beauty Therapy Modelling and Deportment School in Melbourne.

Teacher Janelle Sutton said some of her students went home and teach their parents the basics of good table manners.

'A lot of our young girls don't know how to pick up a knife and fork properly,' she said.

While home is still the best place to learn table manners, schools are being forced to fill the gaps.

Anne Anderson has been a primary teacher for 25 years in Victoria's Western District. She said school camps were often used to teach children the correct way to lay the table and how to use cutlery.

'They learn about different cutlery, where it's placed, what side to put glasses on and putting condiments on the table,' she said.

---

**1** What reasons are given for children today not being skilled enough to properly use cutlery to eat peas?

**2** What are scientists doing to reintroduce the pea onto children's plates?

**3** Why is it important to learn table manners?

**4** What table manners are you aware of? Where did you learn these rules?

## ▶ Red ketchup feels squeeze

Read the newspaper article and then answer the questions that follow it.

GET GOING

## Red ketchup feels squeeze

WE all know that daffodils are yellow and the sky is blue—unless it is summer in England.

Likewise, one of life's little certainties has been that tomato sauce has come in just one colour—red.

That, however, is to change.

Though most parents would agree tomato ketchup is popular enough with their children, Heinz appears to believe its appeal for youngsters needs to be enhanced.

The lurid green variety will fly in the face of 125 years of tradition.

But Heinz insists it tastes the same and will be fortified with vitamin C to encourage parents to buy it.

'We like to think ourselves as an innovative company so in the US trials we carried out with child focus groups and the kids said they wanted the red sauce in a new colour,' said Heinz yesterday.

'Blue was tried out first but deep green was settled on as it was considered to have more "kitchen logic".'

The green ketchup is expected to be a hit on St Patrick's day. Made from green tomatoes, it relies on a splash of 'natural colouring' for its distinctive hue.

'Green is going to be a shocker for a lot of adults, but kids don't have those hang-ups,' said Heinz. 'The core idea is to give children more control and fun over their food.'

Heinz is launching the green ketchup in squeezable bottles in America in October and its success will be closely monitored in Britain.

But the British Nutritional Foundation said: 'Ketchup is high in sugar and shouldn't be smothered on everything. Adding vitamins and minerals to foods that are bad for you can be misleading for parents.'

—DAILY MAIL

1. Why is Heinz creating a green tomato sauce?
2. Would you consume green tomato sauce? Why or why not?
3. What does green tomato sauce taste like?
4. List some of the ingredients found in green tomato sauce.
5. Outline the nutritional concerns about consuming tomato sauce.
6. Create a jingle to promote the sale of green tomato sauce.

# Chapter 2

# Decisions! Decisions!: Deciding What to Eat

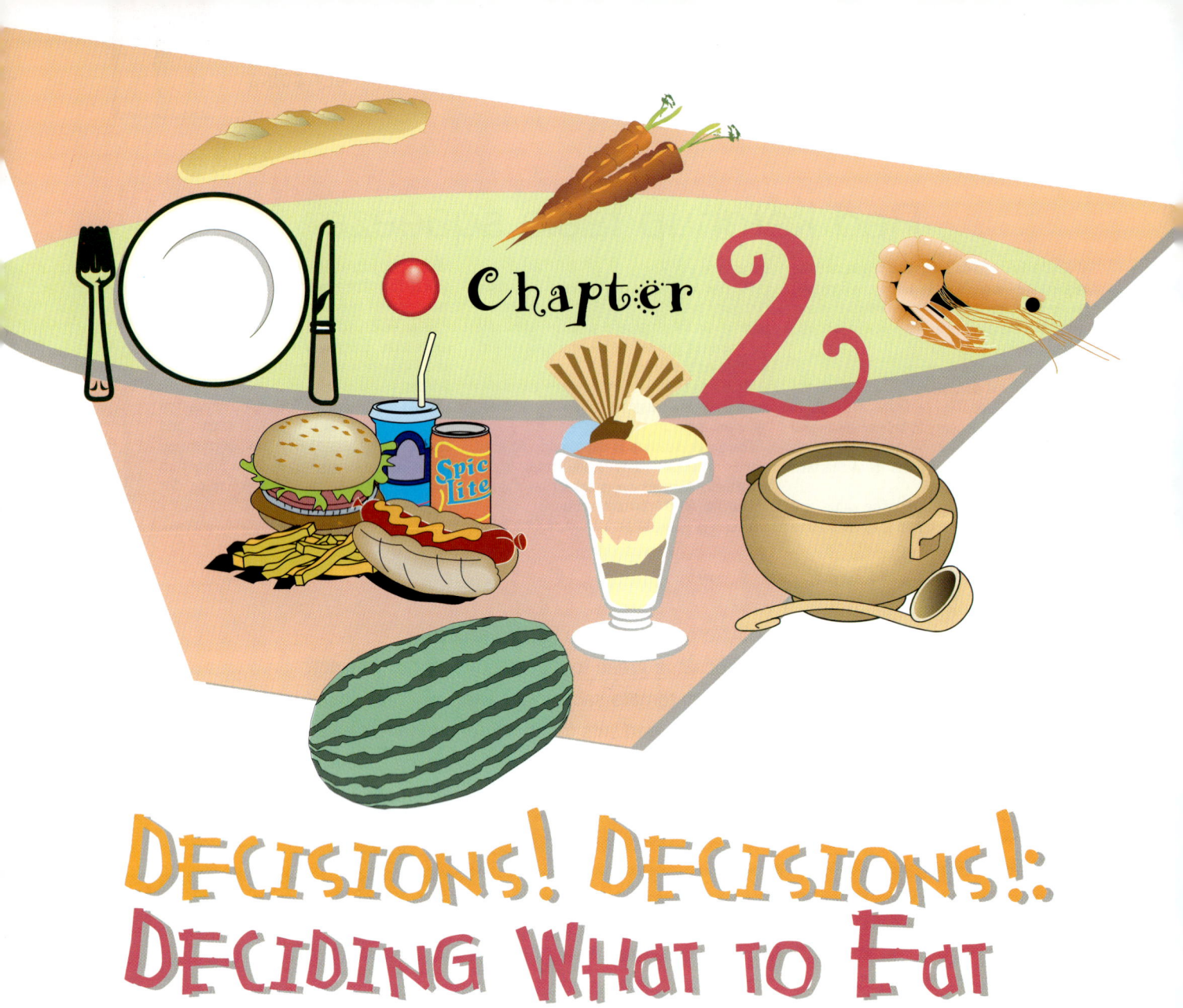

**e-fact**

Did you know that the average person consumes 50 tonnes of food in a lifetime?

We make lots of decisions each day. Many of them relate to what we eat. Food choice is influenced by many factors. These include:
- social factors
- cultural and religious factors
- economic factors
- nutritional knowledge
- media influences

## Social Factors

The family is an important social influence on our food choice. Often we eat a particular food because that is what is served up to us at meal times or what is available in the pantry or fridge.

Today more and more families are leading busier lifestyles. For example, often both parents are working. We frequently say they are 'time

DECISIONS! DECISIONS!

poor' and because of this they tend to buy pre-prepared, convenience and takeaway foods because they are quicker, requiring less time to prepare. Such food tends to cost more.

Recent statistics state that 25 per cent of food expenditure was on takeaway food in 1988; this rose to 33 per cent in 1998. Australia spent $1.1 billion on takeaway food in 1995.

As a result of our busier lifestyles, there has been an increase in such food products as:

- takeaway food or fast food, such as fish and chips, McDonald's and KFC
- pre-prepared food, such as spaghetti sauces, marinated meats or pre-cut vegetables
- home meal replacements, such as frozen dinners

People also use appliances, such as microwaves and sandwich toasters, to reduce the amount of time spent on food preparation

**e-Fact**

The most popular takeaway foods are sandwiches, chips and hamburgers.

We are also influenced by what our friends eat. As we grow up, the influence of peers on our food choice becomes stronger. Adolescents, in particular, will meet their friends at local takeaway shops. They will often select similar food in order to sit and eat together and therefore gain the acceptance of the peer group.

## Cultural and Religious Factors

Our ancestors' countries of origin greatly influence the types of food we eat. This is because we eat food familiar to our eating habits. Today Australia is a very multicultural country: people come from many different countries, such as Britain, Italy, France, Russia, Greece, America, China,

READY, SET, GO!

**e-fact**

Do you believe that beer was often served with breakfast in medieval England?

Japan, Vietnam, India, Africa and the Middle East. These people have influenced our eating habits and significantly changed our diet by bringing with them a range of food and cooking methods. A range of speciality stores now exists to cater for specific cultures.

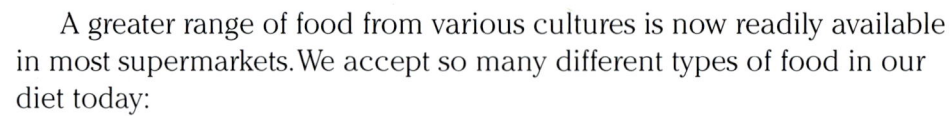

A greater range of food from various cultures is now readily available in most supermarkets. We accept so many different types of food in our diet today:

- Italy: lasagne, spaghetti
- Greece: kebabs, souvlaki
- Mexico: tacos
- China: fried rice
- Japan: sushi
- Russia: beef stroganoff
- France: croissants, baguettes
- America: Coca-Cola, hamburgers
- Sweden: pickled herrings

Religious factors also influence food intake and are closely interwoven with culture (see the table on the next page). Australia is a predominantly Christian country; however many other religions are practised.

DECISIONS! DECISIONS!

| Religion | Food restricted |
|---|---|
| Islam | Pork, alcohol |
| Mormonism (or Church of Jesus Christ of Latter-Day Saints) | Meat, tea, coffee, alcohol |
| Judaism | Shellfish, pork, meat and dairy foods not served at the same meal, meat slaughtered in a specific way |
| Seventh-Day Adventism | Meat, chicken, fish, coffee, tea, alcohol |
| Hinduism | Will not kill any animal for food, but will eat meat from animals slaughtered by members of another religious group |

Religious festivals are important because of the food associated with them. Think about the food you associate with Christmas and Easter. Compare your thoughts with other members of the class.

# Economic Factors

The amount of money available to spend on food will influence food selection:
- Cheaper cuts of meat, like sausages, tend to be higher in fat. Leaner cuts, such as fillet steak, tend to be more expensive.
- Fresh food is not subject to the goods and services tax (GST). For example, a fresh uncooked chicken attracts no tax, but a pre-cooked barbecued chicken does. The fresh chicken requires more time and energy to prepare though. Fresh fruit and vegetables are GST-free.
- Fat-reduced products, such as milk and yoghurt, tend to be more expensive than their full-cream counterparts.

Foods that do not attract GST

Foods that do attract GST

READY, SET, GO!

# Nutritional Knowledge

Often our knowledge about nutrition will influence what we eat. If we played sport, we would eat complex carbohydrates to provide us with energy to perform well. We would even drink water to keep hydrated. Nowadays we are aware of the many health benefits of consuming fish and nuts regularly in our diet; however we did not know this a decade ago. Our knowledge about nutrition is constantly changing. For example, some people used to believe that bread was fattening, but we now know it is the spreads, rather than the bread, that increase the fat content.

Often our nutritional knowledge will allow us to make informed choices about what food to consume.

Reading and understanding labels is important to those people who want to buy nutritious food and maintain a healthy food intake. Today we have over 15 000 different types of food to choose from. A food label must contain the following information:

- the name of the product
- an ingredient list
- a nutritional breakdown of the key nutrients—the food-labelling laws changed on 2 January 2001 so that all food products, except ice cream, jam and meat pies, should have a nutritional panel that provides a breakdown of nutrients, including:
  —energy (kilojoules)
  —protein
  —fat, with saturated fat listed separately
  —carbohydrates, with total sugars listed separately
  —sodium
- the percentages of the key ingredients listed
- the name and address of the manufacturer, packer, importer or vendor
- the country of origin
- identification of where the food was produced—as well as a 'batch' number so that the food can be traced to its packaging plant
- the date of packaging or a use-by date

The table below looks at some of these points in more detail.

**e-fact**
Did you know that a glass of orange juice is equivalent to three oranges, minus the fibre?

**e-fact**
Did you know that thinner potato chips contain more fat than thicker ones? (They have a greater surface area to volume ratio.)

| Food label information | Nutritional knowledge |
| --- | --- |
| Ingredients | Ingredients are listed in decreasing order of weight—the main ingredient comes first, except 'water' as it may be listed at the end. |
| Additives | All approved food additives have a number. It must be listed with its classification, such as emulsifier, food acid, flavour enhancer, mineral salt and thickener. |
| Use-by date | All foods with a shelf-life of less than two years must have a use-by date. Their storage conditions must also be correctly listed. |
| Claims about nutritional value | Food labels cannot make inaccurate or misleading claims, nor should they make such health claims as 'weight reducing'. A label can say 'Calcium is essential for healthy bones' or 'Yoghurt is a good source of calcium', but it cannot say that yoghurt makes strong bones as no one food is responsible for health. Recent changes to food laws have meant that most food labels must provide a panel that outlines nutritional information about the food's protein, fat, carbohydrate, sodium, potassium and kilojoule levels, as well as the amount of any other nutrient for which a claim is being made. |

DECISIONS! DECISIONS!

## Do you know how to read a food label?

# Food Labels
## What Do They Mean?

**ANZFA**
Australia New Zealand Food Authority
TE MANA WHAKARITE KAI MŌ ĀHITEREIRIA ME AOTEAROA

Food labels are a useful source of information. In December 2000, Australian and New Zealand Health Ministers decided that labels on most packaged food would be improved. These changes must all be made by December 2002.

**1. Nutrition labelling.** Until now nutrition labelling has only been compulsory where a food makes a nutrition claim such as 'reduced fat' or 'low salt' or is used for a special purpose such as infant formula or a sports food. The nutrition claim on this yoghurt is 'Good source of calcium'. Detail about the amount of calcium and other nutrients in the yoghurt can be found in the nutrition information panel on the side of the tub. Many, but not all, food manufacturers included this information voluntarily. In future all manufactured foods will have a nutrition information panel so you can make a comparison between them. There are a few exceptions such as very small packages and foods like herbs and spices, tea, coffee and foods sold unpackaged (if a claim is not made) or foods made and packaged at the point of sale.

Nutrition information panels will provide information on the levels of energy (kilojoules), protein, total fat, saturated fat, carbohydrate, sugars and sodium, as well as any other nutrient about which a claim is made on the label. Nutrition information can help you make informed food choices which can lead to better nutrition for you and your family.

**2. Percentage labelling.** Packaged foods will also carry labels which show the percentage of the key or characterising ingredients or components in the food product, if they are present. This will enable you to compare similar products. The characterising ingredient for this strawberry yoghurt is strawberries and you can see from the ingredient list that it has 9% strawberries. An example of a percentage of a characterising component would be the amount of milk fat (a component of milk) in ice cream. In some foods, such as plain bread, there are no characterising ingredients.

**3. Name or description of the food.** Foods must be labelled with an accurate name or description. Labels or descriptions must not mislead consumers, therefore this strawberry yoghurt must contain strawberries.

**4. Food recall information.** Although recalls of unsafe or unsuitable foods are rare, labels must have the name and business address in Australia or New Zealand of the manufacturer or importer, as well as the lot and batch number of the food (or date coding). This makes food recalls more efficient and effective.

**5. More information for allergy sufferers.** The label changes are good news for people with allergies or intolerances to foods. The main foods, food ingredients or components of an ingredient that can cause in some individuals severe adverse reactions - such as peanuts and other nuts, seafood, fish, milk, gluten, eggs and soybeans - must be declared on the label however small the amount. In the case of this yoghurt it is the milk. This declaration is usually in the ingredients list. There must also be an advisory statement on the label where people may be unaware of a possible health risk posed by unpasteurised milk, unpasteurised egg, aspartame, quinine, caffeine in kola beverages and guarana contained in foods and warning statements where people may be unaware of a severe health risk posed by an allergen in a food, for example a warning statement for the bee product, royal jelly, which can cause severe reactions in asthmatics.

**6. Date marking.** Foods with a shelf life of less than two years must have a 'best before' date. It may still be safe to eat those foods after the best before date but they may have lost quality and some nutritional value. Those foods that should not be consumed after a certain date for health and safety reasons, such as infant formula which may be a baby's sole source of food, must have a 'use by' date. An exception is bread which can be labelled with a 'baked on' or 'baked for' date if its shelf life is less than seven days.

Label callouts:
- 6. Date marking. (IMPROVED)
- 8. Labels must tell the truth.
- 10. Legibility requirements.
- 5. More information for allergy sufferers. (IMPROVED)
- 11. Storage requirements.
- 3. Name or description of the food.
- 1. Nutrition labelling. (IMPROVED)
- 2. Percentage labelling. (IMPROVED)
- 7. Ingredients list. (IMPROVED)
- 9. Food additives.
- 12. Country of Origin.
- 4. Food recall information.

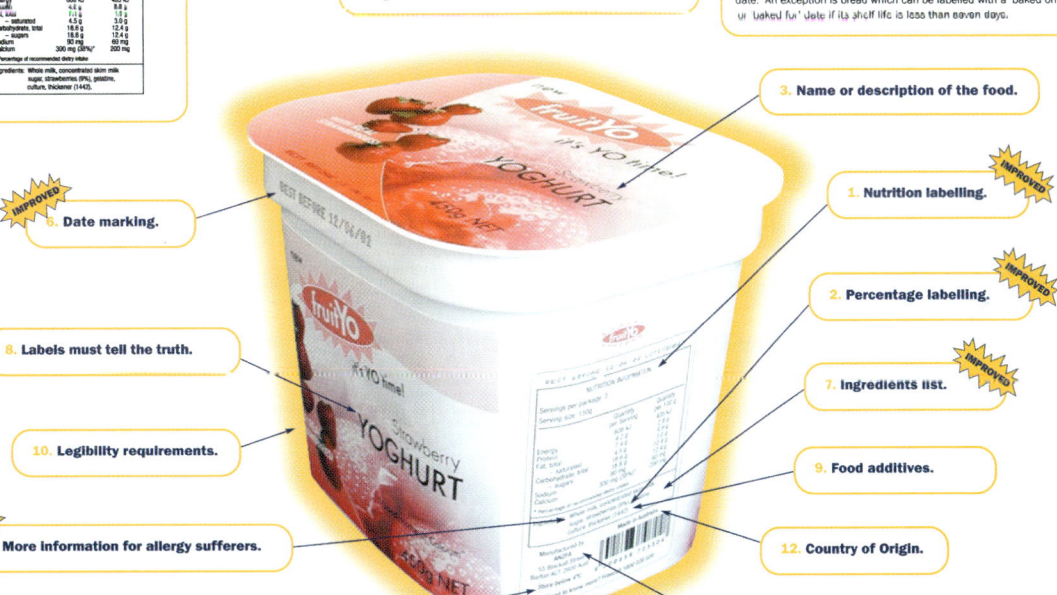

**7. Ingredients list.** Ingredients must be listed from greatest to smallest by ingoing weight including added water. Where there are very small amounts of multi-component ingredients, (under 5%) it is permitted to list the 'composite' ingredient only, for example the chocolate (rather than cocoa, cocoa butter and sugar) in a choc chip cookie or the tomato sauce (rather than tomatoes, capsicum, onions, herbs) on a frozen pizza. This does not apply to any additive or allergen which must be listed however small the amount.

**8. Labels must tell the truth.** Suppliers must label food products with accurate minimum weights and measures information. Weights and measures declarations are regulated by Australian State and Territory and New Zealand Government fair trading agencies. Fair trading laws and food laws in the States and Territories and New Zealand require that labels do not mislead, for example, if the label says it is strawberry yoghurt then it must contain strawberries.

**9. Food additives.** Food additives have many different purposes, including making processed food easier to use or ensuring food is preserved safely. They may come from a synthetic or a natural source. For example, emulsifiers help us spread margarines, and preservatives help to keep food safe or fresh longer. All food additives must have a specific use, must have been assessed and approved by ANZFA for safety and must be used in the lowest possible quantity that will achieve their purpose.

Food additives must be identified, usually by a number, and included in the ingredients list. This allows those people that may be sensitive to food additives to avoid them. A thickener has been used in this yoghurt- its additive number is 1442. A full list of numbers and additives can be obtained from the ANZFA website.

**10. Legibility requirements.** Labels must be legible, with prominent type which is distinct from the background, and in English. The type in legal warning statements must be at least 3mm high, except on very small packages.

**11. Storage requirements.** Where specific storage conditions are required in order for a product to remain safe until its 'Use-by' date, manufacturers must include this information on the label, for example this yoghurt should be kept refrigerated at or below 4°C.

**12. Country of Origin.** The country of origin requirements differ between Australia and New Zealand. In Australia, packaged, and some unpackaged, foods must state the country where the food was made or processed. This could just be identifying the country where the food was packaged for retail sale and, if any of the ingredients do not originate from that country, a statement that the food is made from imported or local and imported ingredients. In New Zealand, country of origin requirements only apply to wines and some cheeses. This matter is currently under review for both countries. Also Australian legislation lays down rules about 'Product of Australia' which means it must be made in Australia from Australian ingredients and 'Made in Australia' which means it is made in Australia with some significant imported ingredients.

**Need more information?** You can contact ANZFA by calling (02) 6271 2222 in Australia or (04) 473 9942 in New Zealand or by emailing info@anzfa.gov.au. The website address is www.anzfa.gov.au or www.anzfa.govt.nz. **For expert nutrition and dietary advice:** Contact your doctor or an accredited practising dietitian. **In Australia:** Visit the 'Find a Dietitian section' of the Dietitians Association of Australia Internet site www.daa.asn.au, check the Yellow Pages or call 1800 812 942 to find an Accredited Practising Dietitian near you. Or contact Nutrition Australia at www.NutritionAustralia.org. **In New Zealand:** Contact the New Zealand Nutrition Foundation on (09) 575 3419 or email nznf@cybernet.co.nz.

**About ANZFA** The Australia New Zealand Food Authority (ANZFA) is an independent bi-national organisation. ANZFA is the result of a partnership between Australia's Commonwealth, State and Territory governments and the New Zealand Government. ANZFA's primary role is to protect the health and safety of the people of Australia and New Zealand by maintaining a safe food supply.

# READY, SET, GO!

18

## Paradise Lites
### 97% FAT FREE
**WHOLEMEAL CRISPBREAD**

- ✓ HIGH FIBRE
- ✓ REDUCED SALT
- ✓ NO CHOLESTEROL

### INGREDIENTS
WHOLEWHEAT FLOUR, GLUCOSE, MALT EXTRACT, SUGAR, VEGETABLE FATS & OILS, MINERAL SALTS (500, 503, 341), YEAST, SALT. MAY CONTAIN TRACES OF DAIRY AND SEED.

### NUTRITION INFORMATION
Servings per package - approx 5
(6 biscuits per serve)

| | per serving 43g | per 100g |
|---|---|---|
| Energy | 663kJ (158 Cal) | 1542kJ (367 Cal) |
| Protein | 5.0g | 11.9g |
| Fat - total | 1.2g | 2.8g |
| - (saturated) | 0.25g | 0.57g |
| - (unsaturated) | 0.95g | 2.23g |
| Cholesterol | Nil | Nil |
| Carbohydrate - total | 32g | 72.7g |
| - sugars | 1.4g | 3.3g |
| Dietary Fibre | 3.0g | 7.0g |
| Sodium | 141mg | 329mg |
| Potassium | 110mg | 257mg |

© & ™ NHF 1988 used under licence

---

## REV
HOMOGENISED, PASTEURISED, LOW FAT MILK
SEEK MEDICAL ADVICE BEFORE USE IN INFANT FEEDING
KEEP REFRIGERATED AT OR BELOW 4°C. STORE IN AN UPRIGHT POSITION

### NUTRITION INFORMATION
Servings per package 8
Serving size 250mL

| | Per Serving (250mL) | Per 100mL |
|---|---|---|
| ENERGY | 536kJ (128 Cal) | 215kJ (51 cal) |
| PROTEIN | 10.3g | 4.1g |
| FAT, total | 3.0g | 1.2g |
| saturated | 2.0g | 0.8g |
| CARBOHYDRATES, total | 14.3g | 5.9g |
| sugars (lactose) | 14.3g | 5.9g |
| SODIUM | 173mg | 69mg |
| CALCIUM (%RDI*) | 375mg (47%) | 150mg (19%) |

*RDI - RECOMMENDED DAILY INTAKE

PRODUCED UNDER LICENCE FROM PARMALAT FOODS AUSTRALIA PTY LTD
842 WELLINGTON ROAD ROWVILLE VIC 3178
© & ™ NHF 1988 USED UNDER LICENCE
Ingredients: low fat milk, skim milk solids.
CONSUMER ENQUIRIES:
FREECALL 1800 676 961
PRODUCT OF AUSTRALIA
PROUDLY PRODUCED IN VICTORIA AND NSW

9 310220 915192

---

9 300601 016921
AUSTRALIA

### Country
These two numbers identify the country of the organisation issuing the number. 93 indicates that the Australian Product Numbering Association (APNA) allocated the numbers.

### Product's manufacturer
These five numbers identify the product's manufacturer. These numbers are allocated by the APNA to its members.

### Specific product
These five numbers are allocated by the manufacturer to each of its specific products.

### Check digit
This number's purpose is to check the accuracy of the reading of the whole number. This check is done by scanning devices.

DECISIONS! DECISIONS!

## Media Influences

**e-fact**
Did you know that Scandinavian law does not allow children's food to be advertised on TV?

Advertisements promote various food products. They inform the consumer about what is available, encourage product competition, and create desires and descriptions of foods that are not always advantageous for good health. For example, chocolate bars are often promoted as providing energy; however this energy is from sugars and not sustained energy from complex carbohydrates. Do chocolate bars really help us to 'work, rest and play'? Advertisements for chips, soft drinks and snack bars frequently promote a fun and carefree lifestyle, but they neglect to reveal their high fat, salt or sugar content.

Advertising does inform us about new products on the market. For instance, we become aware of the range of modified milks available, such as no fat, skim and added-calcium and added-iron varieties.

In the past, primary products—meat, milk, fruit and vegetables—were not widely advertised. This has gradually changed over recent years, but the advertisements of these products are not as extensive as others.

### ▶ TV feeding up little fatties

Read the newspaper article and then answer the questions that follow it.

## TV feeding up little fatties

**By TONY RINDFLEISCH**

Television is making our children fat by studding their favourite shows with advertisements encouraging them to eat fast food.

Research has found that about 80 per cent of the food promoted on TV during children's programs contains little nutritional value.

Ice cream, chocolate, soft drinks, biscuits and hamburgers are the most advertised foods.

All are high in sugar, fat or salt and none is included in Australian Dietary Guidelines for Children.

A study by South Australia's Flinders University has found few advertisements for food for a nutritionally balanced diet during children's viewing times.

Researcher Julie Zuppa found that most of the foods were suitable to eat only occasionally and in small amounts.

She said anxieties about tooth decay, obesity and anaemia were only part of the problem of junk food.

Ms Zuppa said food advertising for children was a major public health issue because habits formed in childhood often were carried into adult life.

She monitored 63 hours of advertisements during children's viewing times on three commercial networks.

Fast foods were advertised most frequently on Saturday mornings—more than nine advertisements an hour.

Ms Zuppa's findings follow a call by one of Australia's foremost nutritionists for a ban on food and confectionery advertising during children's television programs.

Rosemary Stanton said fast food manufacturers were similar to tobacco companies, which, she claimed, aimed their products at children.

High sugar content in foods resulted in high kilojoules which gave children the feeling of being full, meaning they were less likely to seek other foods rich in nutrients.

High-fat diets were directly related to obesity, which could lead to coronary, heart and cardiovascular diseases and diabetes.

Diets high in sodium, which made up 90 per cent of the salt added to food, could contribute to increased blood pressure and hypertension.

1. Why is television making our children fat?
2. What percentage of food advertised during children's programs is of little nutritional value?
3. List foods that are commonly advertised. Why are they not included in the Australian Dietary Guidelines for Children?
4. Where do the foods identified in question 3 belong in *The Australian Guide to Healthy Eating*?

## WEB EXTRAS

**www.foodstandards.gov.au**
The Australia New Zealand Food Authority (ANZFA) is an independent bi-national organisation that aims to protect the health and safety of the people of Australia and New Zealand by maintaining a safe food supply.

**www.cannedfood.org/quality.html**
The Canned Food Information Service supports the Australian canned food industry and the suppliers of tin plate and steel cans.

**www.sanitarium.com.au**
The twin goals of Sanitarium are to provide Australians with healthy foods that actively improve wellbeing and to offer easy-to-understand nutrition information.

## e-Fact

'Making something from scratch' is a North American expression that is becoming popular. It means to make products, such as cakes, biscuits and pancakes, using fresh or raw ingredients with no packet mixes.

---

5  According to the article, what dietary-related diseases may result from consuming the foods identified in question 3?

6  Why are childhood food habits a concern?

7  Why do you think fast foods are more frequently advertised on Saturday mornings?

8  Do you believe that there should be a ban on food and confectionery advertising during children's programs? Why or why not?

9  Outline the concerns raised by Rosemary Stanton about consuming foods with:
  a  high sugar content
  b  high fat content
  c  high salt content

10  Complete your own research. Watch television for one hour. Record the number of food advertisements. How many advertisements were for energy-dense foods—that is, high in fat, salt and sugar and low in fibre? How many were for nutrient-dense foods? What conclusions can you draw from your investigation?

## QUESTIONS

1  Complete these slogans and identify the product being advertised.
  a  Just like a chocolate milkshake, only _____.
  b  The Great Australian _____.
  c  Make Friday night _____ night.
  d  Not too heavy, not too _____.
  e  Flavour you can _____.
  f  Beans means _____.
  g  _____, legendary stuff.

2  Compare no-fat, low-fat and full-cream milk prices. Tabulate or graph your findings. What conclusions can you draw?

3  How much takeaway food did you consume last week? How many meals were made from scratch? How many were pre-prepared, such as pre-cooked chicken, bottled tomato sauces and pre-chopped vegetables?

4  Read the following statements and indicate which factor—social, economic, cultural, religious, media or nutritional information—has the greatest influence.
  a  We always purchase fish and chips on Friday night.
  b  We always eat roast turkey on Christmas Day.
  c  It is quicker to cook in the microwave.
  d  I eat fish on Good Friday.
  e  I use chopsticks to eat.

5  Research the cultural backgrounds of each of the members in the class.

6  Survey the class to determine what each member ate for breakfast that morning. Graph the data. You might like to use a spreadsheet. Then discuss the factors that may have influenced the various foods selected.

DECISIONS! DECISIONS!

**7** Refer to the Safeway labelling website (www.woolworths.com.au/dietinfo/rsa8.asp) and then answer these questions.

    **a** Why is shopping a complicated process today?

    **b** What information has to be provided on food labels?

    **c** How must ingredients be listed on food labels? Are there any exceptions to the rule? Explain.

    **d** Read the ingredient lists for the two breakfast cereals provided below. Which one do you think is the most nutritious? Why?

**NUTRITION INFORMATION**
Servings per package - 36
Serving size - 30 g (3/4 metric cup)†

|  | per 30 g SERVE | Per 30 g WITH 1/2 cup whole milk | per 100 g |
|---|---|---|---|
| ENERGY | 482 KJ (115 Cal) | 853 KJ (203 Cal) | 1607 KJ (383 Cal) |
| PROTEIN | 1.6 g | 6.1 g | 5.4 g |
| FAT | 0.1 g | 5.2 g | 0.3 g |
| CARBOHYDRATE |  |  |  |
| - TOTAL | 26.6 g | 32.8 g | 88.5 g |
| - SUGARS | 11.0 g | 17.2 g | 36.5 g |
| DIETARY FIBRE | 0.4 g | 0.4 g | 1.2 g |
| CHOLESTEROL | 0 mg | 19 mg | 0 mg |
| SODIUM | 169 mg | 238 mg | 564 mg |
| POTASSIUM | 73 mg | 286 mg | 243 mg |
| THIAMIN (VIT B1) | 0.28 mg | 0.35 mg | 0.92 mg |
| (% R.D.I.*) | (25%) | (31%) | (83%) |
| RIBOFLAVIN (VIT B2) | 0.4 mg | 0.6 mg | 1.4 mg |
| (% R.D.I.*) | (25%) | (36%) | (83%) |
| NIACIN | 2.5 mg | 3.3 mg | 8.3 mg |
| (% R.D.I.*) | (25%) | (33%) | (83%) |
| VITAMIN C | 10.0 mg | 11.4 mg | 33.3 mg |
| (% R.D.I.*) | (25%) | (28%) | (83%) |
| FOLATE | 50 µg | 58 µg | 167 µg |
| (% R.D.I.*) | (25%) | (28%) | (83%) |
| CALCIUM | 80 mg | 242 mg | 267 mg |
| (% R.D.I.*) | (10%) | (30%) | (33%) |
| IRON | 3.0 mg | 3.1 mg | 10.0 mg |
| (% R.D.I.*) | (25%) | (25%) | (83%) |
| ZINC | 1.8 mg | 2.3 mg | 6.0 mg |
| (% R.D.I.*) | (15%) | (19%) | (50%) |

\* Recommended Dietary Intake (Aust/NZ)
† Cup measurement is approximate and is only to be used as a guide. If you have any specific dietary requirements please weigh your serving.

**INGREDIENTS:** WHOLE RICE, SUGAR, COCOA POWDER, SALT, SKIM MILK POWDER, WHEY POWDER, MINERALS (CALCIUM CARBONATE, IRON, ZINC OXIDE), MALT EXTRACT, FLAVOUR, DEXTROSE, VITAMINS (VITAMIN C, NIACIN, RIBOFLAVIN, THIAMIN, FOLATE), MAY CONTAIN TRACES OF PEANUTS AND/OR OTHER NUTS.

**INGREDIENTS:** RIZ COMPLET, SUCRE, CACAO EN POUDRE, SEL, LAIT ÉCRÉMÉ EN POUDRE, LACTOSÉRUM EN POUDRE, MINÉRAUX PARFUM, DEXTROSE, VITAMINES (VITAMINE C, NIACINE, RIBOFLAVINE, THIAMINE, FOLATE), PEUT CONTENIR DES TRACES DE CACAHUÈTES ET D'AUTRES NOIX, NOISETTES, ETC.

**KELLOGG'S - MADE IN AUSTRALIA, EXPORTING TO THE WORLD.**

**NUTRITION INFORMATION**
Servings per package - 11
Serving size - 45 g (3/4 metric cup)†

|  | per 45 g | Per 45 g WITH 1/2 cup skim milk | per 100 g |
|---|---|---|---|
| ENERGY | 700 KJ (169 Cal) | 904 KJ (215 Cal) | 1574 KJ (375 Cal) |
| PROTEIN | 3.7 g | 8.3 g | 8.3 g |
| FAT | 0.7 g | 0.9 g | 1.6 g |
| CARBOHYDRATE |  |  |  |
| - TOTAL | 36.4 g | 43.1 g | 80.8 g |
| - SUGARS | 10.0 g | 16.5 g | 22.3 g |
| DIETARY FIBRE | 3.8 g | 3.8 g | 8.4 g |
| CHOLESTEROL | 0 mg | 5 mg | 0 mg |
| SODIUM | 22 mg | 92 mg | 49 mg |
| POTASSIUM | 192 mg | 409 mg | 427 mg |
| THIAMIN (VIT B1) | 0.28 mg | 0.33 mg | 0.61 mg |
| (% R.D.I.*) | (25%) | (29%) | (56%) |
| RIBOFLAVIN (VIT B2) | 0.4 mg | 0.6 mg | 0.9 mg |
| (% R.D.I.*) | (25%) | (33%) | (56%) |
| NIACIN | 2.5 mg | 3.3 mg | 5.6 mg |
| (% R.D.I.*) | (25%) | (33%) | (56%) |
| FOLATE | 100 µg | 108 µg | 222 µg |
| (% R.D.I.*) | (50%) | (53%) | (111%) |
| IRON | 3.0 mg | 3.0 mg | 6.7 mg |
| (% R.D.I.*) | (25%) | (25%) | (83%) |

\* Recommended Dietary Intake (Aust/NZ)
† Cup measurement is approximate and is only to be used as a guide. If you have any specific dietary requirements please weigh your serving.

**INGREDIENTS:** CEREALS (WHOLE WHEAT, ROLLED OATS, RYE), SULTANAS, RAW SUGAR, DRIED APRICOT PIECES (DRIED APRICOT, FRUCTOSE, GLYCEROL, THICKENER [1422], MALTODEXTRIN, VEGETABLE OIL, SOYA BEAN FLOUR, NATURAL FLAVOUR, FOOD ACID [CITRIC ACID]), MALT EXTRACT, SALT, HONEY, CASTER SUGAR, VITAMINS (NIACIN, RIBOFLAVIN, FOLATE, THIAMIN), MINERAL (IRON), MAY CONTAIN TRACES OF PEANUTS AND/OR OTHER NUTS.

    **e** Who decides what additives can be used in food in Australia?

    **f** Why is it important to list food additives?

    **g** List the fourteen different types of food additives. Explain what their functions are.

    **h** Construct a table like the one below, identifying the name and type of the following additives: 422, 1204, 904, 421, 171, 297, 276, 925, 900 and 356.

| Number | Name | Type |
|---|---|---|
|  |  |  |

READY, SET, GO!

    i  What do use-by dates mean and why is it important to list storage conditions?
    j  In terms of nutritional value what is required on a food label?
    k  In terms of nutritional value what is not allowed on a food label?
    l  Explain why the following nutritional claims are permitted to appear on food labels:
—no added sugar
—gluten-free
—no added salt
—fibre increased
—low joule
—no cholesterol or cholesterol-free
—fat reduced or low fat
—no artificial additives, colours or flavourings
Find food labels that make each of the above nutritional claims.

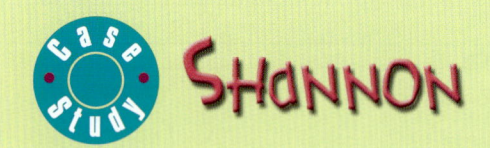

## Case Study: Shannon

Shannon is thirteen years old. She lives with her mother and ten-year-old brother, Kane, in a small country town. Shannon loves riding her horse and playing tennis on the weekend. She is in Year 7 and her favourite subjects are science, physical education and drama. She also enjoys 'chatting' on the computer with her friends and watching television. Shannon rides her bike to school and is responsible for looking after Kane every school night. Their mother works as a pharmacist, so they are responsible for many household tasks, such as unloading the dishwasher and feeding their pet cat. Shannon and Kane also help prepare dinner most nights, though every Friday night they order in takeaway because their mother works late—Shannon's favourite is pizza. Saturday is shopping day. Shannon and Kane always help their mother by selecting what foods they want to have in their school lunches. Their mother encourages them to read labels before buying foods.

### Question

Identify the factors that may influence Shannon's food intake.

# Chapter 3
# Supermodels: Discover Healthy Eating Models

To help Australians make healthy food choices, different food models have been developed over recent years. At primary school, you might have learnt about the five food groups or the healthy eating pyramid. One of the most recent food models is *The Australian Guide to Healthy Eating*, which was launched in 1998 by the Commonwealth Department of Health and Family Services. The foods in *The Australian Guide to Healthy Eating* have been used to form the structure of this book and will be examined in more detail throughout each chapter. First, let us look at the Healthy Eating Pyramid and the Dietary Guidelines for Australians.

READY, SET, GO!

# The Healthy Eating Pyramid

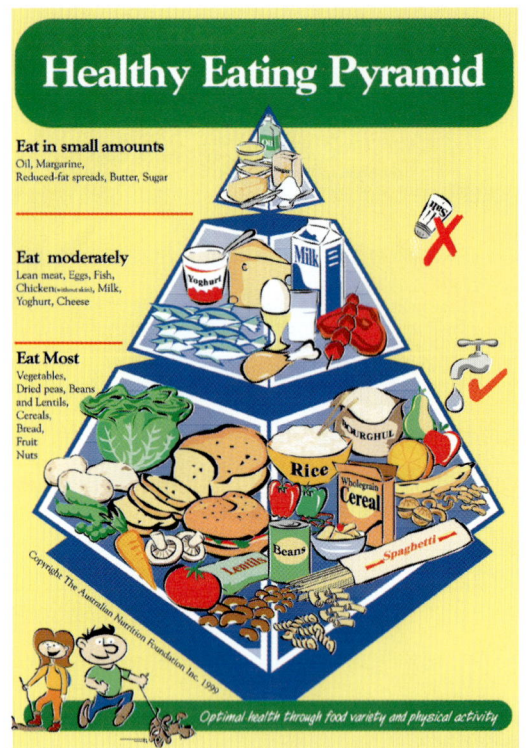

Nutrition Australia developed the Healthy Eating Pyramid. It provides a guide as to the proportions of food we should eat and suggests which food we should mostly eat, which we should eat in moderation and which we should eat in small amounts.

The biggest part of the pyramid is the 'eat most' section. It contains such plant foods as fruit, vegetables, bread, cereals, dried peas and beans, legumes and nuts, all of which are high in fibre and complex carbohydrates and low in fat.

The middle section of the Healthy Eating Pyramid contains animal foods that are high in protein, including lean meat, fish, eggs, poultry (without skin), milk, yoghurt and cheese. We should eat a moderate amount of these foods.

The top section of the pyramid contains food we should eat in small amounts. It includes butter, oil, margarine, sugar and reduced-fat spreads.

The Healthy Eating Pyramid also recommends we should drink water and not add salt to our food.

A range of eating pyramids has been developed to suit specific nutritional needs. For example, Nutrition Australia has also developed a Healthy Eating Pyramid for Vegetarians, specifically lacto-ovo vegetarians—they do not eat meat; however they do consume eggs and milk.

Numerous eating pyramids have also been developed to meet the nutritional needs of other cultures. These include the Mediterranean, Mexican, Indian, Russian and Asian eating pyramids.

Healthy Eating Pyramid for Vegetarians

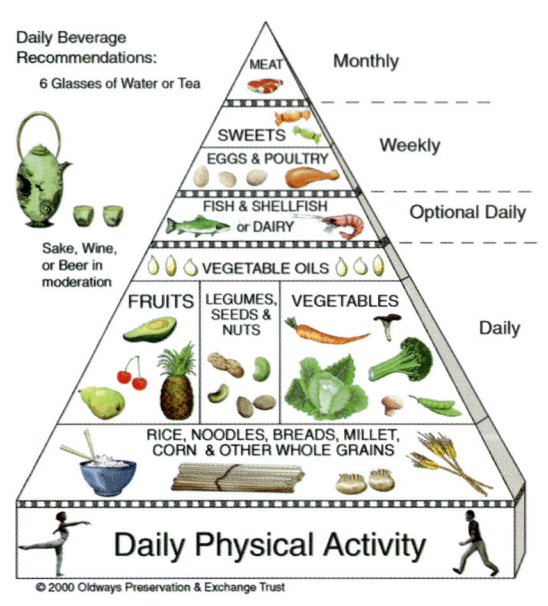

Traditional Healthy Asian Diet Pyramid

### ▶ Food I like to eat pyramid

Draw your own food pyramid according to the food you like to eat. What food would you put in your 'eat most' section? What do you like to eat in moderation and what do you like to eat in small amounts?

Now compare your food I like to eat pyramid with the Australian Healthy Eating Pyramid. How different are they? Does your pyramid need modification? How?

# Dietary Guidelines for Australians

The National Health and Medical Research Council (NHMRC) developed the Dietary Guidelines for Australians to assist people with making healthy food choices. In addition, the guidelines make suggestions about physical activity, breastfeeding and alcohol consumption—in other words, they suggest that it is not only the food we consume that is important but also our lifestyle choices.

## Dietary Guidelines for Australians (1992)

1. Enjoy a wide variety of nutritious foods.
2. Eat plenty of breads and cereals (preferably whole grain), vegetables (including legumes) and fruits.
3. Eat a diet low in fat and, in particular, low in saturated fat.
4. Maintain a healthy body weight by balancing physical activity and food intake.
5. If you drink alcohol, limit your intake.
6. Eat only a moderate amount of sugars and foods containing added sugars.
7. Choose low salt foods and use salt sparingly.
8. Encourage and support breastfeeding.

### Guidelines on specific nutrients

1. Eat foods containing calcium. This is particularly important for girls and women.
2. Eat foods containing iron. This applies particularly to girls, women, vegetarians and athletes.

Source: NHMRC Dietary Guidelines for Australians (1992) (http://www.health.gov.au/hfs/nhmrc/advice/diet.htm)

## Dietary Guidelines for Children & Adolescents

- Encourage and support breastfeeding.
- Children need appropriate food and physical activity to grow and develop normally. Growth should be checked regularly.
- Enjoy a wide variety of nutritious foods.
- Eat plenty of breads and cereals, vegetables (including legumes) and fruits.
- Low-fat diets are not suitable for young children. For older children, a diet low in fat and, in particular, low in saturated fat is appropriate.
- Encourage water as a drink. Alchohol is not recommended for children.
- Eat only moderate amounts of sugars and foods containing added sugars.
- Choose low-salt foods.

### Guidelines on specific nutrients

- Eat foods containing calcium.
- Eat foods containing iron.

Source: NHMRC Dietary Guidelines (1995) (http://www.health.gov.au/hfs/nhmrc/advice/diet.htm)

# The Australian Guide to Healthy Eating

One of the first food models to be developed was the Five Food Groups, which has been incorporated into *The Australian Guide to Healthy Eating*. Look at the diagram that shows *The Australian Guide to Healthy Eating* and you will notice there are five different-sized sections. You will probably also notice these sections are the same as those in the Five Food Groups model. The relative size for each indicates how much we should eat, so in this sense it is similar to the Healthy Eating Pyramid. Check out the tables on pages 27–28 to see how many serves from each section is recommended.

You will notice that the table showing sample serves has two different amounts for each age group. The sample serves shown in pink are for those of you who are of small or average size with a less active lifestyle. The sample serves in yellow are for those of you who are more active and are an average or large size.

SUPERMODELS

## THE AUSTRALIAN GUIDE TO HEALTHY EATING
## Enjoy a variety of foods every day

*Vegetables, legumes*

*Fruit*

*Milk, yoghurt, cheese*

*Lean meat, fish, poultry, eggs, nuts, legumes*

Drink plenty of water

*Bread, cereals, rice, pasta, noodles*

Choose these sometimes or in small amounts

**Suggested sample serves for children and adolescents from *The Australian Guide to Healthy Eating***

| Children and adolescents | Bread, cereals, rice, pasta and noodles | Vegetables and legumes | Fruit | Milk, yoghurt and cheese | Meat, fish, poultry, eggs, nuts and legumes | Extra foods |
|---|---|---|---|---|---|---|
| 4–7 years | 5–7 | 2 | 1 | 2 | $\frac{1}{2}$ | 1–2 |
|  | 3–4 | 4 | 2 | 3 | $\frac{1}{2}$–1 | 1–2 |
| 8–11 years | 6–9 | 3 | 1 | 2 | 1 | 1–2 |
|  | 4–6 | 4–5 | 1–2 | 3 | 1–1$\frac{1}{2}$ | 1–2 |
| 12–18 years | 5–11 | 4 | 3 | 3 | 1 | 1–3 |
|  | 4–7 | 5–9 | 3–4 | 3–5 | 1–2 | 1–3 |

READY, SET, GO!

**Suggested sample serves for women from** *The Australian Guide to Healthy Eating*

| Women | Bread, cereals, rice, pasta and noodles | Vegetables and legumes | Fruit | Milk, yoghurt and cheese | Meat, fish, poultry, eggs, nuts and legumes | Extra foods |
|---|---|---|---|---|---|---|
| 19–60 years | 4–9 | 5 | 2 | 2 | 1 | 0–2$\frac{1}{2}$ |
|  | 4–6 | 4–7 | 2–3 | 2–3 | 1–1$\frac{1}{2}$ | 0–2$\frac{1}{2}$ |
| Pregnant | 4–6 | 5–6 | 4 | 2 | 1$\frac{1}{2}$ | 0–2$\frac{1}{2}$ |
| Breastfeeding | 5–7 | 7 | 5 | 2 | 2 | 0–2$\frac{1}{2}$ |
| 60+ years | 4–7 | 5 | 2 | 2 | 1 | 0–2 |
|  | 3–5 | 4–6 | 2–3 | 2–3 | 1–1$\frac{1}{2}$ | 0–2 |

**Suggested sample serves for men from** *The Australian Guide to Healthy Eating*

| Men | Bread, cereals, rice, pasta and noodles | Vegetables and legumes | Fruit | Milk, yoghurt and cheese | Meat, fish, poultry, eggs, nuts and legumes | Extra foods |
|---|---|---|---|---|---|---|
| 19–60 years | 6–12 | 5 | 2 | 2 | 1 | 0–3 |
|  | 5–7 | 6–8 | 3–4 | 2–4 | 1$\frac{1}{2}$–2 | 0–3 |
| 60+ years | 4–9 | 5 | 2 | 2 | 1 | 0–2$\frac{1}{2}$ |
|  | 4–6 | 4–7 | 2–3 | 2–3 | 1–1$\frac{1}{2}$ | 0–2$\frac{1}{2}$ |

*The Australian Guide to Healthy Eating* also says to drink plenty of water and to be aware of foods that we should only choose to eat either sometimes or in small amounts—such as biscuits, soft drinks, chocolate, ice cream, lollies—and that are high in fat—such as pies, sausage rolls, pasties, potato chips and pastries.

## ▶ *e*-Food models

www.health.gov.au/pubhlth/strateg/food/guide/index.htm

You can find out more about *The Australian Guide to Healthy Eating* by visiting this website. Information packages are also available from the Commonwealth Department of Health and Family Services.

1. Why has *The Australian Guide to Healthy Eating* been developed?
2. According to *The Australian Guide to Healthy Eating*, what are the recommendations for eating a healthy diet?
3. What factors can influence the foods we choose?
4. Why are some of the foods pictured outside the five sections?
5. In which two sections are legumes listed? Why?
6. Explain what is meant by 'sample serves' for each group.
7. Explain why the sample serves vary within each age category. For example, why does it say that twelve to eighteen year olds need either four or five to nine serves of vegetables or legumes?

SUPERMODELS

**8** What suggestions are made for vegetarians in relation to choices from the meat, poultry, fish, eggs, nuts and legumes group?

www.dhs.vic.gov.au/nphp/signal/whatis.htm

The Strategic Inter-Governmental Nutrition Alliance (SIGNAL) website provides information about an action plan aimed at improving the nutritional health of Australians. SIGNAL's key priority is to develop and implement the 'Eat well Australia: A national framework for action in public health nutrition, 2000–2010' strategy.

1. Where do the representatives of SIGNAL come from?
2. What does the 'Eat well Australia' strategy focus on?
3. Why is a national approach important?

www.nal.usda.gov/fnic/etext/000023.html

The US Department of Agriculture (USDA) has a Food and Nutrition Information Center (FNIC) located within the National Agriculture Library (NAL). Within the FNIC is an index of different ethnic/cultural food guide pyramids. Their website is really great fun and provides interesting comparisons about the nutritional needs of a range of ethnic/cultural groups.

# Ready, Set, Go!: Assessment Task

This assessment task addresses the outcome TEMA0501 from the Technology Key Learning Area. The website www.olliesworld.com/aus/index.html, which is entitled Ollie's Recycles, can be used as a basis to complete it.

## ▶ Packaging

1. What is packaging and what do we use it for?
2. List three ways to decrease the amount of packaging.
3. 
   a. What materials may packaging be made from?
   b. List the different types of materials used for packaging in the pantry, at either home or school.
   c. Why do you think so many different types of materials are used for packaging?

## ▶ The Three R's of Recycling

### Reuse

1. What is composting? Outline the difference between aerobic and anaerobic composting.
2. What essential ingredients are required for composting? Ensure you list some 'green' and 'brown' organic matter.

# READY, SET, GO!

## WebExtras

**www.cansmart.org**
The Steel Can Recycling in Australia website comes care of the Steel Can Recycling Council and Planet Ark. It provides information both on how and why steel cans should be recycled and on the history of the steel can.

**www.gould.edu.au/wastewise/kids/intro.htm**
The Waste Wise for Kids website contains many different activities and ideas about recycling, composting, litter and wormeries.

**www.gould.edu.au/waste_stop/act_intro.htm#contents**
The Waste Stoppers website provides a series of information sheets written by the Gould League of Victoria, with funding from EcoRecycle Victoria, that deal with all aspects of waste minimisation.

**www.ecorecycle.vic.gov.au**
EcoRecycle Victoria aims to bring about significant changes in the way Victoria addresses resource recovery, recycling and waste management.

**www.moove.com.au**
The Moove website provides lots of fun activities designed to raise awareness about the benefits of dairy products and of recycling.

**www.3rcentre.evoblue.com**
This website contains information on the 3rCentre—a resource centre providing a range of products made from recycled materials and promoting sustainability.

**3** Draw a flow chart to outline how you could create a compost.
**4** What is a wormery?
**5** Define the term *castings*.

## Recycle

### Steel

**1** What items are made from steel?
**2** Why is steel used for packaging?
**3** Draw the four symbols that are used to indicate that an item is made from steel.
**4** Describe how steel is recycled.

### Aluminium

**1** List the uses of aluminium.
**2** Draw the symbol that is used to indicate that cans are made from aluminium.
**3** Besides the symbol, how else can you tell whether a can is made from steel or aluminium?
**4** The symbol used to represent the recycling of aluminium is a continuous circle. Draw your own interpretation of this symbol and explain why it is a circle.

### Plastic

**1** What do we use plastic for?
**2** Outline the advantages of plastic.
**3** There are three types of recyclable plastic bottles. Draw up the table below in your workbook and complete it.

| Plastic Bottles ||||
|---|---|---|---|
| Type of recyclable plastic | Symbol | Characteristics of bottle | What it is recycled into |

### Glass

**1** Draw the recycled glass symbol.
**2** What are the colours of recycled glass?
**3** Why should we encourage glass recycling?
**4** What determined the colour of glass?
**5** Why is it important to separate the different colours of glass when recycling?

### Paper

**1** What is paper used for?
**2** List the two types of paper that are recycled.
**3** Describe how paper is recycled.

## Reduce

**1** Describe ways you could reduce the amount of waste.
**2** Why is reducing waste important, especially in regard to landfill?

## Section 2

# Cereals, Bread, Rice, Pasta and Noodles

### The Australian Guide to Healthy Eating sample serve

2 slices of bread

1 medium bread roll

1 cup cooked rice, pasta, noodles

1 cup porridge, 1$\frac{1}{3}$ cups breakfast cereal flakes or $\frac{1}{2}$ cup muesli

$\frac{1}{3}$ cup flour

# Chapter 4

# Let's Get Cereals: Discover Grains

**e-RIDDLE**
Q: Why did the pancake play baseball?
A: He was a good batter.

What do you think of when you hear the word *cereal*? Do you think of breakfast cereals or cereal grains like wheat, oats, barley or rye? Perhaps you think of such cereal products as muesli bars, bread, rice bubbles or flour?

In the early days of civilisation, people found their food by hunting and gathering. Then came the agricultural revolution, when people learnt how to grow and harvest their own crops. Some of the earliest grown were cereals—the edible seeds of grasses.

*Cereal* is derived from the Latin *Cerealia munera*, meaning 'gift from Ceres', the Roman goddess of the Earth and harvest. The early origins of this term reflect its importance to early civilisations.

LET'S GET CEREALS

33

# Classify Cereals

We generally classify cereals according to the type of grain. In Australia, wheat is the major cereal crop, accounting for approximately 90 per cent of all grain production. It is also the staple grain for much of the world's population. Even in Asian countries, where rice is the staple grain, wheat is used to make noodles and steamed bread. The other 10 per cent of grain production in Australia is primarily made up of maize (corn), barley, oats and rice.

# Properties of Cereals

The nutritional value of all cereal grains is similar, providing us with carbohydrates, protein, fibre, vitamins and minerals.

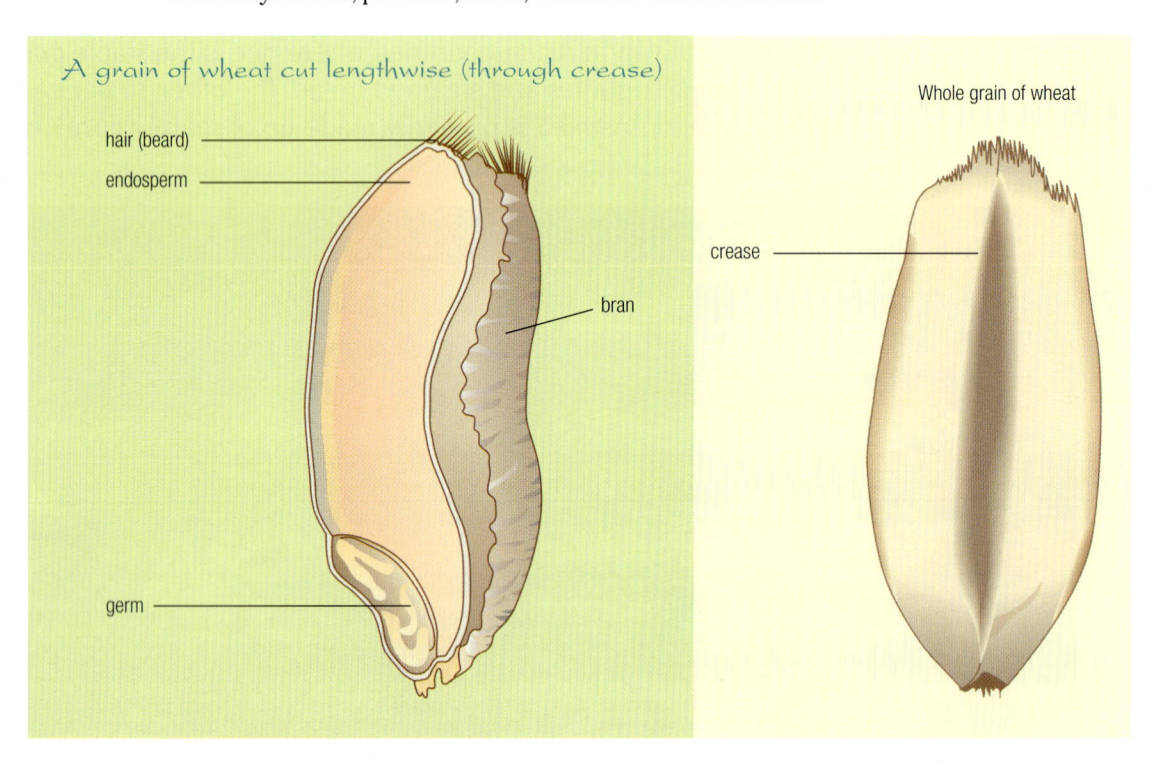

*A grain of wheat cut lengthwise (through crease)*

- hair (beard)
- endosperm
- bran
- germ

Whole grain of wheat
- crease

Cereals contain around 70 per cent carbohydrates, which are an important source of energy. This is why eating bread or cereals for breakfast is an essential start to the day. We usually associate protein with meat and fish and often forget that it is also found in cereals, especially wheat and oats, which are good sources. The main vitamins found in cereals are the B group vitamins and vitamin E. Minerals include iron, magnesium and zinc.

# Focus on Fibre

Fibre is found in the outer layers of cereal grains, as well as other plant foods. There are two types of fibre: soluble and insoluble fibre. Soluble fibre slows down the rate of digestion. Insoluble fibre cannot be broken down by our digestive system, helps to prevent constipation and keeps our bowel regular, removing waste quickly.

Information has been discovered recently about resistant starch, which is referred to by some as a type of fibre. Resistant starch is undigested when it reaches the bowel and therefore promotes the growth of good bacteria, which also help keep the bowel healthy. It can be found in porridge oats and wholegrain bread.

LET'S GET CEREALS

## WEB EXTRAS

**www.awb.com.au**
AWB is the major national grain-marketing organisation in Australia.

**www.uncletobys.com.au**
The Uncle Tobys website contains product and nutritional information.

**www.gograins.grdc.com.au**
Go Grains is a nutrition communication initiative established for the Australian grains industry. It provides information on the nutrition and health benefits of grains and pulses.

*The Australian Guide to Healthy Eating* suggests eating a wide variety of grains, including more wholegrain products. Insoluble fibre should form a major part of our daily food intake because it can pass undigested through the stomach and small intestine. The properties of fibre are thought to help prevent health problems, such as obesity, diabetes, some cancers and heart disease. Eating a high-fibre diet also helps us to reduce our hunger and makes us feel more full; high-fibre foods also tend to be low in fat. The value of low-fat foods is highlighted in the chapter 21, 'Fat Is Not a Four-letter Word'.

A good guide to the amount of fibre in grams that children should consume is their age plus five. For example, a ten year old should consume 10 + 5—that is, 15 grams of fibre. How many grams should you be consuming each day?

While one slice of white bread contains 0.8 gram of fibre, one slice of wholegrain bread contains 2.8 grams of fibre. One 30-gram bowl of Cornflakes contains 0.8 gram of fibre; an equivalent bowl of Coco Pops contains 0.4 gram of fibre and two Weet-bix contain 3.2 grams of fibre. Do you think you eat enough fibre every day?

## Cereal Products

There are a large number of products available made from cereal grains. We discovered earlier that wheat is the largest grown cereal crop in Australia. It is mainly used to produce flour, which is then used to make such products as breads, pastries, pizza bases, pasta and a range of baked goods, like muffins, cakes and biscuits.

When cereal products are produced, the cereal grain is processed and sometimes part of it is removed. The outer or bran layers of the cereal grain contain fibre and when they are removed, so too is the fibre. Products such as refined flour have had the bran and the endosperm removed, with the flour being produced mainly from the starchy endosperm. For this reason, it is vital that we try to eat wholegrain products—made from the whole grain, such as wholegrain bread and brown rice.

CEREALS, BREAD, RICE, PASTA AND NOODLES

## Questions

1. Make a list of all the kinds of information you can find on the Uncle Tobys Apricot Fruesli Bars packet on the next page.
2. Who produces Uncle Tobys Apricot Fruesli Bars?
3. If you ate an apricot fruesli bar at recess, what percentage of your recommended daily intake (RDI) of dietary fibre would you have consumed? (Hint: Refer to the section on fibre to work out how much you should eat each day.)
4. Who do you think is the target market for this product? In other words, what groups of people would this product appeal to?
5. When the same manufacturer produces similar products, the food industry refers to these products as line-extensions. Name two line-extensions mentioned on the packet.
6. The packet is made out of 'Australian Recycled Cartonboard', indicating that it is 80 per cent recycled. Why do you think the manufacturer puts this on the packet?
7. Uncle Tobys works with Coastcare groups.
    a  What do you think both groups are aiming to achieve?
    b  Why do you think this information is important to consumers?
8. Indicated on the packet is 'Trade Mark Pending'. What does this mean?
9. Write a paragraph about why you think the symbol 'Proudly Australian' can be found on the packet.
10. What is the value to consumers of the 'Health Watch—"On the run"' section?

## ▶ Let's remember

1. Define the word *cereal*.
2. What is the origin of *cereal*?
3. What is the main cereal grown in Australia?
4. List three other cereal grains grown in Australia.
5. What is the name of the plant that produces corn?
6. Which nutrient in cereals provides us with energy?
7. Explain the different types of fibre.
8. What is resistant starch? Explain.
9. List three advantages of a high-fibre diet.
10. How many grams of fibre are found in two Weet-bix?

LET'S GET CEREALS

37

## UNCLE TOBYS breakfree Apricot Fruesli Bars

**97% FAT FREE**

- Made with real fruit
- Goodness of 4 grains
- Good source of dietary fibre
- Low in fat

Reward your body and your taste buds with other Uncle Tobys breakfree snacks: Oven Baked Twists, Oven Baked Fruit Bars, Fruit & Yoghurt Bars, and Cookies.

6 BARS 270g

---

Uncle Tobys breakfree snacks are quality products from Goodman Fielder Ltd

### NUTRITION INFORMATION

SERVINGS PER PACKAGE: 6
SERVING SIZE: 1 BAR (45g)

| | QUANTITY PER SERVING | QUANTITY PER 100g |
|---|---|---|
| ENERGY | 650kJ | 1450kJ |
| PROTEIN | 2.3g | 5.0g |
| FAT, TOTAL | 1.4g | 3.0g |
| - SATURATED | 0.3g | 0.6g |
| CARBOHYDRATE, TOTAL | 34.7g | 77.2g |
| - SUGARS | 10.5g | 23.4g |
| DIETARY FIBRE | 3.1g | 6.8g |
| SODIUM | 35mg | 80mg |
| POTASSIUM | 160mg | 350mg |

### INGREDIENTS

ROLLED WHEAT, GLUCOSE SYRUP (FROM WHEAT), ROLLED TRITICALE, APRICOTS (9%), CEREAL PUFFS (RICE, OATS, WHOLE WHEAT), HI-MAIZE™ BRAN ('HI-MAIZE' CORNFLOUR, WHEAT BRAN, WHEATGERM, CASTOR SUGAR, SOY FIBRE, GOLDEN SYRUP, MINERAL [CALCIUM CARBONATE], SALT, FLAVOUR [VANILLA], EMULSIFIER [471], VITAMINS [B1, B2, NIACIN, FOLATE, F]), UNCLE TOBYS OATS, DEXTROSE (FROM WHEAT), FRUCTOSE, SUGAR, HUMECTANTS (GLYCERINE, SORBITOL), LITE CRISPIES (RICE, WHOLE WHEAT, WHEAT GERM, WHEAT FLOUR, SUGAR, MALT EXTRACT, SALT, MINERALS [CALCIUM, IRON], FLAVOUR, EMULSIFIER [471], VITAMINS [B1, B2, NIACIN, FOLATE]), HONEY, SUNOLA VEGETABLE OIL¹, WHEAT MALTODEXTRIN, NATURAL EMULSIFIER (LECITHIN), VEGETABLE GUM (PECTIN), FLAVOUR, FOOD ACID (CITRIC), PRESERVATIVE (220)*. MAY CONTAIN TRACES OF PEANUTS AND OTHER NUTS.

* SOME DRIED FRUITS CONTAIN SULPHUR DIOXIDE TO MAINTAIN NATURAL COLOUR AND SHELF LIFE.

CEREALS, BREAD, RICE, PASTA AND NOODLES

### ▶ Let's investigate

1. How many different brands of breakfast cereals can you think of?
2. Collect a range of breakfast cereal packets, with each student bringing in a different packet from home. Identify all of the different types of vitamins and minerals found in the breakfast cereals.
3. Using the breakfast cereal packets collected in question 2, complete a bar graph to show the difference in fibre content between the different types of breakfast cereals. Which breakfast cereals would you consider to be healthy? Explain.
4. Complete the table on the left by including as many examples of cereal products as you can think of for each of the cereal grains. One example for each grain has been provided for you.

| Cereal grain | Cereal product |
| --- | --- |
| Wheat | Wheat flour |
| Corn | Cornmeal |
| Rice | Rice flour |
| Oats | Porridge oats |
| Barley | Pearl barley |

### ▶ Puzzled

#### Cereal scramble
Unjumble the following types of cereals and cereal products.

1. souccsou
2. gulbur
3. delsisn
4. yelbar
5. telnopa
6. ridgepor
7. teamgerwh
8. rabn
9. toas
10. norc

#### Which grain?
1. Pumpernickel bread is made from _____.
2. Triticale is a cross between _____ and _____.
3. The cereal used in muesli is _____.
4. Burgul is sometimes referred to as cracked _____.
5. Tortillas are made from _____.

#### Grain match
Match each of the grains—oats, wheat, barley, rice and corn—with these food products.

1. Anzac biscuits
2. country vegetable soup
3. sushi
4. tacos
5. gnocchi

### ▶ @-Cereal

**www.kellogg.com.au**

The Kellogg's website promotes the importance of breakfast. Follow the links to breakfast through the nutrition and lifestyle button and then read the reasons why breakfast is important. Write a couple of paragraphs to justify why we should all eat breakfast.

**www.choice.com.au/cp/food/fibre1.cfm**

Go to the *Choice* magazine website and do the quiz 'Are you getting enough fibre?'

**www.choice.com.au/cp/food/cereals1.cfm**

Find a breakfast cereal to meet your needs by entering your request and searching the *Choice* database.

LET'S GET CEREALS

# Let's Produce

## Ingredients

- 15 grams margarine
- 1/8 red capsicum
- 1/4 onion, finely chopped
- 1/2 teaspoon curry powder
- 3/4 cup vegetable stock
- 2/3 cup couscous
- 3 tablespoons corn kernels
- 2 tablespoons sultanas
- 2 teaspoons of parsley, chopped finely

### Vegetarian curry couscous  (serves 2)

**Method**

1. Melt margarine. Add capsicum and onion and cook until soft.
2. Add curry powder. Stir.
3. Add stock and bring to the boil.
4. Add couscous. Remove from heat, cover and stand for five minutes.
5. Add corn, sultanas and parsley.
6. Serve immediately.

## Ingredients

- 1 egg
- 1 cup buttermilk
- 1 cup flour
- 1 tablespoon baking powder
- pinch salt
- 2 teaspoons sugar
- 2 tablespoons melted butter
- 1/4 teaspoon vanilla (optional)
- 1/3 cup blueberries
- 4 tablespoons maple syrup

### American hotcakes  (serves 2)

**Method**

1. Beat egg and buttermilk together.
2. Mix flour, baking powder, salt and sugar together.
3. Add melted butter and vanilla to the egg and milk mixture.
4. Add moist ingredients to dry ingredients. Mix well to remove lumps.
5. Cook in a non-stick, lightly greased frypan.
6. Serve with blueberries amd maple syrup.

**e-HINT**

Use a 1/3 measuring cup to pour the batter in the pan. Cook the hotcakes on one side until bubbles appear and then flip and cook on the other side. (The first side of the pancake always takes longer to cook than the flip side!)

## Ingredients

- 1 cup frozen raspberries
- 2 cups self-raising flour
- 1/2 cup sugar
- 1/2 cup chocolate bits
- 1 lightly beaten egg
- 60 grams melted butter
- 1 1/4 cups buttermilk

### Chocolate raspberry muffins  (makes 12)

**Method**

1. Preheat oven to 180°C.
2. Grease muffin pans.
3. Thaw raspberries.
4. Combine the sifted flour, sugar and chocolate in a bowl.
5. Mix egg, melted butter and buttermilk.
6. Combine the dry and wet ingredients and lightly mix.
7. Gently fold in raspberries.
8. Spoon into pans and bake for approximately fifteen to twenty minutes.

CEREALS, BREAD, RICE, PASTA AND NOODLES

## Mexican nachos (serves 2)

### Ingredients
½ avocado
½ teaspoon lemon juice
3 drops Tabasco sauce
100 grams nachos chips
½ can Mexican chilli beans
75 grams tasty cheese, grated
2 tablespoons sour cream

### Method
1. Mash avocado flesh. Add lemon juice and Tabasco sauce.
2. Place nachos chips in an oven-proof dish. Spread Mexican chilli bean mixture over chips.
3. Sprinkle with cheese.
4. Bake in a moderate oven (180°C) for ten to fifteen minutes, or until cheese is melted and bean mixture is heated through.
5. Place avocado on top of bean mixture. Place sour cream next to avocado.

## Vegetarian triangles (serves 2)

### Ingredients
15 grams butter
¼ onion
¼ teaspoon curry powder
¼ carrot, grated
¼ potato, grated
¼ zucchini, grated
1 tablespoon corn kernels
1 tablespoon bean shoots
1 sheet puff pastry

### Method
1. Melt butter and sauté finely chopped onion for one to two minutes.
2. Add curry powder and cook for one minute.
3. Add carrot, potato, zucchini, corn kernels and bean shoots to mixture. Sauté for a further minute.
4. Cut pastry into four even squares.
5. Divide mixture into four and place filling into middle of pastry. Brush edges of pastry with water. Join edges to form a triangle and pinch together across the join.
6. Brush pastry with melted butter. Place in an oven (200°C) and cook for fifteen minutes or until brown.

### Variation
Use filo or shortcrust pastry instead of puff pastry and make vegetable parcels. Fold one sheet in half and place filling at one end of the pastry. Fold in the sides and roll into a parcel.

**e-HINT**
Prick each pastry with a fork to allow steam to escape and not cause the seams to open.

# Chapter 5

# D'oh!: Discover Bread

**e-RIDDLE**

Q: Why did the bread go to the doctor?
A: He was feeling crumby.

Did you know bread was first made nearly 12 000 years ago? It was most likely roughly crushed grain that was mixed with water and either left in the sun to dry, heated on hot stones or baked by being covered with hot ashes. Consequently, the bread was flat and hard. These first breads were made from the seeds of wild grasses, such as wheat, barley or oats. Later, two stones were used to crush these seeds to make the flour that was used to make bread.

It is claimed that the ancient Egyptians were the first people to invent leavened bread, which has a leavening (or raising) agent that causes it to rise. Apparently, when the Egyptians left their dough to dry in the open air, wild yeast spores landed in it and produced a gas called carbon dioxide. This gas added volume to the dough, making it the first leavened bread.

In the past, the poor consumed brown bread, whereas the rich ate white bread. The ancient Egyptians gave brown bread to their slaves. In England in the eighteenth century, white flour was considered so desirable that bakers would add various ingredients, such as ground-up bones, chalk

CEREALS, BREAD, RICE, PASTA AND NOODLES

## e-fact

The creation of the modern-day pizza coincided with the visit of Italy's Queen Margherita to Naples in 1889. It represented the Italian flag because its three ingredients were red tomato, white mozzarella and green basil.

and wood ashes, to flour to make it look even whiter. It would not have tasted that great! Fortunately, King George III set an excellent example for healthy eating by always selecting brown bread for his lunch. People would cry out his nickname 'Brown George' as his carriage travelled by.

Australia's first settlers brought their own flour with them from England: 448 barrels to last for the first two years of the colony. Bread was a really important part of their diet, so most of their flour was used to make it. After two years, however, very little flour was left, so the settlers used to boil the flour with green vegetables to make it go further. Several times a week, they took their flour to the public bakehouse; a day later they received their bread. When the colony grew and families became established, the girls were taught how to make their own bread.

## e-fact

On average, each Australian consumes about 48 kilograms of bread per year.

## Properties of Bread

Bread is a nutritious food that provides a good source of carbohydrates, protein, fibre and a range of vitamins and minerals. Wholegrain bread is more nutritious than white bread because it contains more fibre and B group vitamins. However all bread is good for you and it is a great snack. You can eat it as a sandwich, toasted, with dips or just by itself.

*carbohydrates* — *protein* — *fibre* — *vitamins & minerals*

# Classify Bread

All over the world there are many different varieties of bread; in many countries it is considered to be a staple food. This means that bread is a regular and important part of the daily diet, and that it plays a significant role in many cultures and religions.

| Type of bread | Description | Diagram |
|---|---|---|
| Bagel | Bagels have a hole in the middle and are made by boiling the bread dough before baking. They are chewy and dense. Many Jewish people eat bagels because they traditionally have no eggs, milk or fat. | |
| Baguette | The baguette is the king of French bread. It is made from wheat flour and is crusty on the outside and soft on the inside. It must be eaten as fresh as possible for the melt-in-the-mouth taste. | |
| Ciabatta | Ciabatta means 'slipper' in Italian. This bread has a flat, oval shape that is slightly crunchy on the outside and soft and chewy inside. | |
| Damper | Damper originates from the Irish, who baked quick bread in an iron pot over an open fire. New settlers in Australia made similar bread and called it damper. | |
| Focaccia | Focaccia originated in the villages of France and the Italian Riviera. It has a flat, rectangular or oval shape and is flavoured with ingredients according to the district where it is made. It is brushed with oil. | |
| Lavash | Lavash are rectangular sheets of flat bread that are about 0.2 centimetre thick with a chewy texture. It dates back to biblical times and was originally made by the Babylonians and Assyrians. | |
| Naan | Naan is a traditional Indian flat bread. It is slightly leavened with yeast and shaped like a teardrop. It is best eaten warm. | |
| Pita bread | Pita is a two-layered flat, oval- or round-shaped bread that virtually has no crumbs. It is also known as pocket bread because it can be split to take fillings. | |
| Soda bread | Soda bread originates from Ireland. It contains no yeast and is leavened with baking soda, which was previously combined with an acid, usually buttermilk. | |
| Soy and linseed bread | Soy and linseed bread has recently become popular owing to research that shows the benefits of eating a diet high in soy and linseed, especially for women. The soy contains isoflavones that contribute to a healthy heart; the linseed provides omega-3 fatty acids that are important for healthy arteries. | |

CEREALS, BREAD, RICE, PASTA AND NOODLES

## Bread Products

### Wonder White

Wonder White is a high-fibre white bread made from unique maize grown in Australia. This maize is ground so finely that the fibre in the flour cannot be seen when made into bread—it is invisible in fact. Consequently, this light and fluffy bread appeals to both parents and children. Wonder White has twice the fibre normally found in white bread. Fibre is important because it ensures that food passes regularly through the body. Wonder White is also high in resistant starch, which is also important because it helps prevent constipation and bowel cancer. Have you ever eaten Wonder White?

For more information on fibre, refer to 'Focus on Fibre' in chapter 4, 'Let's Get Cereals'.

**e-fact**

Soluble fibre is found in grains such as barley and oats. It helps to control blood sugar levels. This is why soluble fibre is particularly beneficial to diabetics. Soluble fibre lowers the risk of heart disease because it assists with lowering blood cholesterol.

Although white bread was always relatively low in fibre, Wonder White bread, which is high in soluble fibre, is a great way for people who prefer white bread to increase their fibre intake.

## Case Study: BREADMAKERS ON THE RISE!

Home-made bread has become quite popular with the invention of breadmakers. Depending on the model, they enable people to make a broad range of white, wholemeal, rye, gluten-free, yeast-free and sweet breads, and possibly dough for bread rolls, pizzas, focaccias, hot cross buns, strudels and other baked products.

The major allure of breadmakers is that they produce freshly baked bread with little effort and that they enable people to experiment with various textures and flavours, such as sun-dried tomato and capsicum or olives.

A breadmaker with a timer program means that you can wake to the aroma of freshly baked bread that you can eat warm for breakfast!

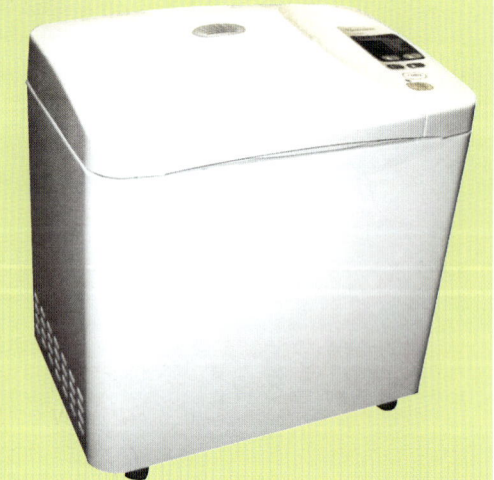

### QUESTIONS

1. Why do you think breadmakers have become popular?
2. What types of breads can be produced by using a breadmaker?
3. **Design** a type of bread you would like to produce with a breadmaker.
4. Survey the class to determine how many people have access to a breadmaker. What types of breads are produced? What is the most common?

D'OH!

## ▶ Let's remember

1. From what grains were the first types of bread made?
2. List three ways the first types of breads were cooked.
3. How were the grains crushed?
4. What is leavened bread?
5. How did the bakers in the eighteenth century make white bread look whiter?
6. Why is bread nutritious?
7. Why is fibre important in our diet?
8. Why is wholegrain bread more nutritious than white bread?
9. Why is bread considered to be a staple food?
10. How many serves of bread are recommended daily?

> **e-DEFINE**
> An isoflavone is a naturally occurring compound present in chickpeas and legumes, such as soy, which has the most concentrated amount.

## ▶ Let's investigate

1. Visit your local bakery. Make a list of the types of breads being sold.
2. Crumby's Bakeries have decided to develop a new product called Damper Delight. What ingredients do you suggest could be added to a basic damper recipe to create Damper Delight?
3. In 1810 bakers could not sell bread unless it was one day old. **Investigate** a current food law related to bread.
4. Think of as many types of breads as possible that begin with the letter *P*.
5. In Australia most flour is made from the wheat grain. Find out what other grains can be used to make flour.

> **e-DEFINE**
> Omega-3 fatty acids are unsaturated fatty acids that are present mainly in fish oils.

## ▶ e-Bread

www.wonderwhite.com.au

1. Click on the 'Twice the facts' section.
   a. List the nutrients found in Wonder White bread. Use the websites of ANZFA (www.foodstandards.gov.au) or Safeway (www.woolworths.com.au/recipes/foodaddtives) to find out what the additive codes stand for.
   b. What is Hi-Maize starch?
2. Click on 'Twice the fibre'.
   a. Why do you think Wonder White was developed?
   b. Why do we need fibre in our diet?
   c. Why is resistant starch important?
   d. Graph the fibre content of regular white and Wonder White bread.
3. Click on 'Twice the food'. Read through the recipes for Wonder White and create your own.
4. Click on 'Twice the fun'. Watch the Wonder White TV advertisement, with Gus and Freddie on the hunt for twice the invisible fibre in Wonder White bread. Write your own promotional advertisement for one of the Wonder White products.

CEREALS, BREAD, RICE, PASTA AND NOODLES

**www.woolworths.com.au/rsa4.asp**

1. Provide two reasons why bread consumption has decreased.
2. What is glycogen?
3. Why would you lose weight if you avoided carbohydrates? How easily do you regain this weight?
4. Is bread 'fattening'?
5. What is the function of carbohydrates?
6. List the nutrients found in bread.
7. Outline the importance of fibre.
8. What is starch? What does recent research say about its importance?
9. List and describe the different bread varieties.
10. Outline how you would store bread.

## ▶ Puzzled

### Bread mix

Match the breads in the first column of the table with their countries in the second.

| Bread | Country |
| --- | --- |
| Baguette | Australia |
| Chapatti | Germany |
| Naan | Mexico |
| Tortilla | Scotland |
| Pumpernickel | France |
| Damper | France |
| Crumpet | India |
| Croissant | India |
| Bap | Britain |

### Superstitions

Read and discuss the following superstitions associated with bread. Do you believe any of them?

1. The soil was often watered with the blood of human sacrifice at the time of sowing a crop of corn.
2. People would try to prove their innocence during the Middle Ages by eating dry bread and not choking.
3. If you threw bread into a fire, you were said to feed the devil.
4. Eating too much bread will give you a hairy chest.
5. Eating crusts will make your hair go curly.

### Bread jumble

Unjumble the following types of breads.

1. yre
2. lleeamhwo
3. gabeetut
4. dosa
5. pamred
6. ccfcaiao
7. mudpapap
8. siarni
9. kneelppmucri
10. prisc

## WEBExtras

**www.brumbys.com.au**
Brumby's mission is to bring the best bread recipes from all around the world and deliver them freshly baked to local suburbs.

**www.bakersdelight.com.au**
Bakers Delight aims to provide delightful bread for their customers, though not without keeping in mind the social and environmental effects of their actions.

**www.hi-maize.com**
The Hi-maize website provides detailed information for health professionals, food companies and consumers.

**www.middleeastbakeries.com.au**
The mission of Middle East Bakeries is 'the best ingredients, the best products, the happiest customers'.

**www.tiptop.com.au**
One of Tip Top's aims is to introduce breads that provide increased value to Australian consumers, such as nutritional and health benefits.

D'OH!

# Let's Produce

## Bruschetta (serves 2)

**Ingredients**
½ 30-centimetre loaf Turkish bread
3 tablespoons olive oil
2 cloves garlic, crushed
3 large ripe tomatoes, diced
black pepper
1 tablespoon fresh basil, finely chopped

### Method
1. Preheat oven to 180°C.
2. Cut the Turkish bread in half, lengthwise.
3. Cut the bread into 10-centimetre pieces.
4. Combine two tablespoons of oil and garlic and brush both sides of the Turkish bread with the mixture.
5. Place the bread on a tray, with the cut side facing upwards.
6. Cook until golden (about ten to fifteen minutes).
7. Transfer to a cake cooler.
8. Place the tomato, remaining oil, basil and pepper in a bowl and gently toss to combine.
9. Place the tomato mixture on bread pieces and serve.

### Variations
**Investigate** other ingredients you could use and **design** a topping for bruschetta. **Investigate** what other breads could be used instead of Turkish bread.

## Damper (serves 1)

**Ingredients**
1 tablespoon butter
2 cups self-raising flour
250 millilitres milk

### Method
1. Preheat oven to 220°C.
2. Rub butter into flour.
3. Add milk and mix into a soft dough.
4. Turn onto floured board and mould into a large, round shape.
5. Mark eight wedges on top and place on tray. Glaze top with milk.
6. Bake for fifteen to twenty minutes or until golden brown.
7. Break into wedges for serving.

### Variations
- For cheesy damper, add three-quarters of a cup of grated tasty cheese and one-quarter of a tablespoon of mustard to the flour mixture.

CEREALS, BREAD, RICE, PASTA AND NOODLES

- For fruity damper, add half a cup of dried fruit and one tablespoon of caster sugar to the flour mixture.
- For a damper roll, roll out the damper dough into a 30 x 20-centimetre shape and spread with two tablespoons of tomato paste, two tablespoons of grated tasty cheese and two tablespoons of sliced ham. Roll up like a Swiss roll and cut into twelve slices.

Investigate and design other ingredients you could use in damper. Some suggestions could include sun-dried tomatoes, herbs, olives, onion and pumpkin. Produce and evaluate your damper ☺ ☺ ☹. Remember to give it a creative name.

## Calzone (serves 2)

### Dough ingredients
- 7 grams dried yeast
- ½ teaspoon sugar
- 2 tablespoons warm water
- 2 tablespoons warm milk
- ¾ cup plain flour
- 1 tablespoon olive oil
- ½ egg, lightly beaten

### Filling ingredients
Select from:
- tomato paste
- tomatoes, sliced
- cheese, diced
- capsicum, diced
- mushrooms, sliced
- ham, cut into strips
- salami, sliced
- pineapple pieces
- onion, diced
- oregano
- thyme
- basil
- black pepper

### Method
1. Place yeast, sugar, water and milk in a bowl and leave in a warm place until mixture becomes frothy.
2. Sift flour and make a well.
3. Add frothy liquid ingredients, egg and oil and knead on a floured surface for about five minutes or until smooth.
4. Cover and rest dough in a warm place for thirty minutes.
5. In the meantime prepare your filling ingredients.
6. Divide dough into two balls and roll into thin pancakes.
7. Spread each round with tomato paste, and place the remaining ingredients on one half of the circle.
8. Brush edges with water and fold over the dough. Pinch down the edges to seal.
9. Place on baking paper on a tray in a preheated 200°C oven and cook for about twenty minutes.

### Variations
Investigate and design your own calzone filling ingredients. Produce and evaluate ☺ ☺ ☹.

D'OH!

## Cheesy crumbed chicken strips (serves 2)

**Ingredients**
- 2 chicken breast fillets
- 1 egg, lightly beaten
- ½ cup plain flour
- ½ cup dry breadcrumbs
- ¼ cup Parmesan cheese, grated
- 1 tablespoon oil
- 30 grams butter
- mixed lettuce
- tomato chunks

### Method

1. Flatten chicken fillets and slice into strips.
2. Place egg and flour into two separate bowls.
3. Mix breadcrumbs and cheese together in another bowl.
4. Dip each strip into flour and then egg and roll in breadcrumb mixture.
5. Heat oil and butter in a frypan and cook the chicken strips until lightly brown.
6. Drain on absorbent paper.
7. Serve on a bed of mixed lettuce and tomatoes.

## Chocolate bread and butter pudding (serves 2)

**Ingredients**
- 1 teaspoon cocoa
- ¾ cup milk
- 50 grams cooking chocolate, roughly chopped
- ½ teaspoon vanilla
- 1 teaspoon brown sugar
- 2 eggs, lightly beaten
- 2 slices thick white bread, buttered on both sides
- 1 tablespoon sultanas
- icing sugar
- cream, ice cream or yoghurt

### Method

1. Blend cocoa with a little of the milk.
2. Add the remaining milk, chocolate, vanilla and sugar and place into a saucepan.
3. Heat gently, stirring constantly until chocolate melts and mixture is smooth.
4. Cool slightly and add egg using a whisk.
5. Using a large scone cutter, cut each slice of bread into a circle shape to fit into the base of a greased soufflé dish.
6. Place circle of bread into soufflé dish and sprinkle with sultanas.
7. Pour chocolate mixture over bread and sultanas.
8. Bake at 170°C for twenty minutes or until set.
9. Sprinkle with icing sugar and serve with cream, ice cream or yoghurt.

## QUESTIONS

1. Why did you mix the cocoa with a little milk before adding the remainder of the milk?
2. Why do you need to cool the chocolate mixture before adding the egg?
3. **Investigate** other ingredients could you use instead of sultanas.
4. **Evaluate** the appearance, flavour, texture and aroma of the pudding ☺ ☻ ☹.

# Chapter 6

# On the Boil: Discover Rice

**e-fact**

A person in Burma consumes, on average, 226.8 kilograms of rice per year. Now that's a lot of rice!

Rice is considered to be one of the oldest cultivated crops. Records show that rice was cultivated in China in 5000 BC; the earliest writings in Thailand, Syria and Egypt also mention it. According to ancient legend, the Emperor of China and his sons used to plant the first rice seeds to mark the beginning of the planting season.

ON THE BOIL

**51**

### e-fact
Approximately 66 per cent of the world's population eat rice daily and over 50 per cent eat rice regularly.

Rice is a small, oval-shaped grain that is grown in warm climates, where there is at least 100 centimetres of rain per year. It is grown in over 100 countries, including Australia; however nearly 90 per cent of all rice produced comes from Asia.

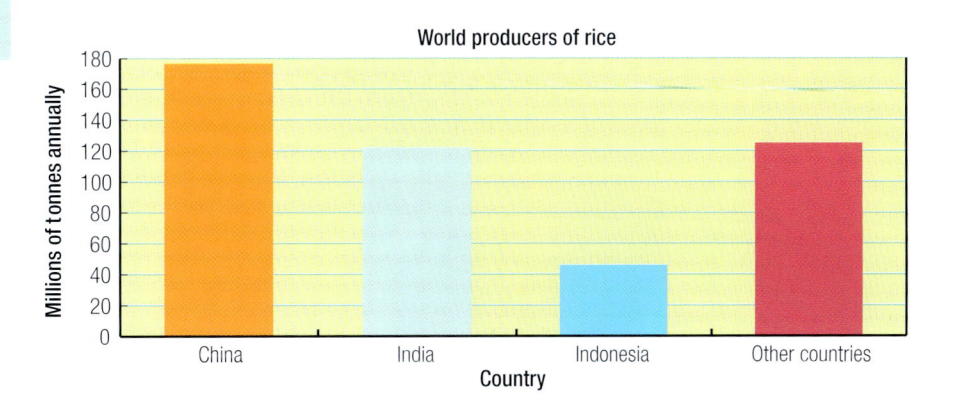

*World producers of rice* (bar chart: China ~175, India ~122, Indonesia ~45, Other countries ~125 millions of tonnes annually)

Rice is an important part of the diets of millions of people, especially those in developing countries: it is a vital source of nutrients. How often do you eat rice?

## How Is Rice Processed?

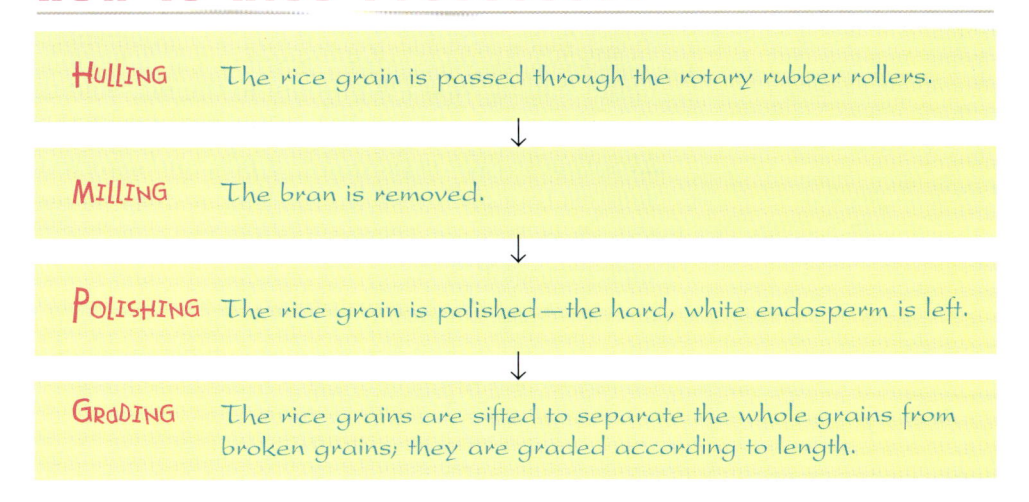

|  |  |
|---|---|
| **Hulling** | The rice grain is passed through the rotary rubber rollers. |
| **Milling** | The bran is removed. |
| **Polishing** | The rice grain is polished—the hard, white endosperm is left. |
| **Grading** | The rice grains are sifted to separate the whole grains from broken grains; they are graded according to length. |

CEREALS, BREAD, RICE, PASTA AND NOODLES

**e-fact**

Did you know that Twisties contain rice?

Rice and their products are used in a range of food. Examples include:
- breakfast cereals, such as Rice Bubbles and Coco Pops
- noodles
- crispbreads
- rice cereal for baby foods
- rice crackers
- rice pasta
- rice biscuits, such as Sakata
- instant cooked rice

## Cooking Rice

Rice is cooked in a variety of ways. All rice-eating people have differing views on how it should be cooked. Some wash rice prior to cooking; others do not. Iranians do; they may even soak it overnight. Italians neither soak nor wash their rice. It really depends on the type of rice being used and the way it is being cooked. The Chinese wash their rice prior to cooking to remove the excess starch, so when the rice is cooked it will have a dry texture with separated grains.

Rice consists of starch grains found in the endosperm. When placed in liquid, these grains swell and absorb the liquid, becoming soft and fluffy. Rice will usually triple its volume when cooked.

Some ways to cook rice include:
- Rapid boiling: Water is brought to the boil and the rice is added, left uncovered and simmered until soft and fluffy.
- Absorption: Rice is placed in a pan and covered with cold water. The water is brought slowly to the boil and simmered until all of it is absorbed. The rice is usually rinsed before cooking to stop the grains sticking together.
- Microwaving: Washed rice is placed into a microwaveable container and covered with water. A lid is placed on top of the container. This method takes about the same time as boiling.

ON THE BOIL

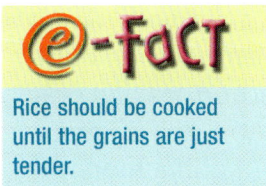

**e-fact**
Rice should be cooked until the grains are just tender.

- Rice cookers: Rice is placed in the bottom of the cooker and covered with enough water so that when you place your fingertip on top of the rice, the level of the water will come to the first joint of your finger.

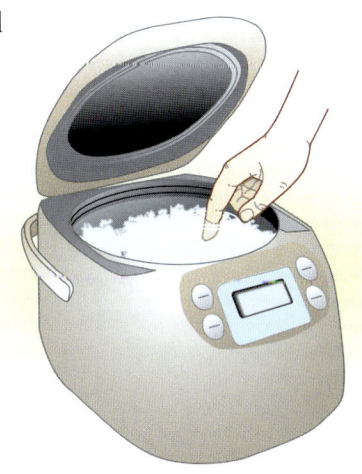

# Properties of Rice

Rice is an excellent source of carbohydrate. Brown rice also provides fibre and the B group vitamins. Below is a table that illustrates the differences between white and brown rice (100 grams uncooked).

| Nutrient | White rice | Brown rice |
|---|---|---|
| Protein (g) | 6.6 | 7.7 |
| Fat (g) | 0.5 | 2.4 |
| Carbohydrate (g) | 79.1 | 77.4 |
| Fibre (g) | 0.8 | 3.2 |
| Iron (mg) | 0.7 | 1.2 |
| Zinc (mg) | 1.1 | 2.1 |
| Thiamin (mg) | 0.08 | 0.35 |
| Riboflavin (mg) | 0.02 | 0.05 |
| Niacin (mg) | 2 | 4.5 |

# Classify Rice

Rice can be either brown or white. The edible rice grain is found in a hard shell-like hull, which is surrounded by several layers of bran.

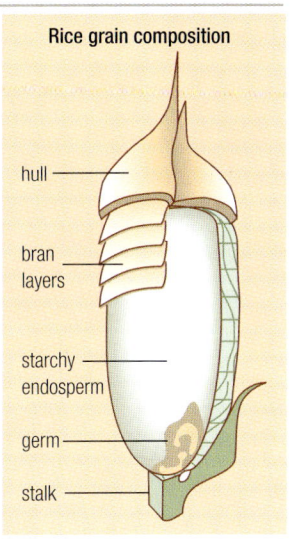

Rice grain composition
- hull
- bran layers
- starchy endosperm
- germ
- stalk

CEREALS, BREAD, RICE, PASTA AND NOODLES

## e-fact

**Wild rice is not really a type of rice; rather it is a long, thin grain that is indigenous to North America.**

Brown rice is unprocessed and contains the whole grain, with most of the bran layers still intact. In contrast, white rice has been polished during the milling process—its bran and germ have been removed, leaving only the endosperm. Thus brown rice is more nutritious and offers a different flavour in comparison to white rice: it has a nutty flavour and a more chewy texture. Brown rice takes longer to cook than white rice because it still has most of its bran in one piece. In fact, it takes about forty minutes to cook, whereas white rice only takes twelve minutes!

The Chinese prefer to eat white rice; they are most likely influenced by Confucius, a fifth-century BC philosopher, who claimed that rice should be as white as possible. Confucius obviously had little nutritional knowledge! The husks from rice grains were removed and most likely sold for polishing precious gems.

Rice is categorised according to the length of its grain: long, medium and short. Long-grain rice tends to stay separated when cooked and usually has a fluffy texture. It is typically used in Indian and Chinese cooking for savoury food. Its grains are about five times as long as they are wide. Examples of long-grain rice include basmati and jasmine rice.

Short-grain rice is oval in shape and produces firm grains that tend to cling together after cooking, making it easier to eat when using chopsticks. Short-grain varieties are used in the West for sweet dishes, such as rice pudding. The Spanish use it in paella, which is a savoury dish that contains fish. The Japanese use a short-grain rice called koshihikari in sushi. Another example of a short-grain rice is Arborio, which is produced in Italy to make risotto. It absorbs liquid during the cooking process, resulting in plump and tender grains with a very creamy texture.

Approximately 1.2 million tonnes of rice is produced each year in Australia, about 80 per cent of which is medium-grain. This type of rice tends to be rounder in shape in comparison to long-grain rice and has a moist texture when cooked. Medium-grain rice, such as calrose, is typically used in savoury dishes.

Beliefs associated with rice:
- It is considered bad luck to spill rice or knock over a rice bowl.
- Rice is thrown at weddings as a sign of life and fertility. Flinging rice at newly-weds indicates a wish that they will be blessed with lots of children!
- In many languages the word for rice is the same as that for food as it was the only source of food.
- Rice is so important in Japan that many shrines are dedicated to the rice god Inari.
- Rice was a measure of wealth for many centuries; it was often used instead of money. For example, the Japanese Samurai were paid in rice.
- No girl would be regarded as eligible for marriage in Java unless she could prepare a good bowl of rice.

ON THE BOIL

## Focus on Sakata Rice Snacks

The Sato brothers started making these rice snacks over fifty years ago in the Japanese village of Sakata, which is located in the fertile rice-growing area known as Shonai. The rice produced there was well known for its superior quality, flavour and fine texture. The Sato brothers have worked tirelessly to produce the thin rice cracker with a fine, crispy texture known as the *usi-yaki* rice cracker. Today, it comes in over 100 varieties.

### QUESTIONS

1. What varieties of Sakata Rice Snacks have you consumed?
2. Write down ten other varieties of rice crackers. Be creative!

## Rice Products

**e-fact**

In Australia, parboiled rice is sold as 'rice that cooks in the fridge'. It has been subjected to steam prior to milling.

Various products made from rice are:
- white rice
- wholemeal rice
- rice flour
- rice bran
- parboiled (quick-cooking) rice
- rice bran
- rice chips
- rice cakes
- rice noodles

### ▶ Let's remember

1. Describe the appearance of a rice grain.
2. Outline the type of climate necessary to grow rice.
3. How much rice do the following countries produce?
   a. China
   b. India
   c. Indonesia
4. What is polished rice?
5. What causes rice grains to become soft and fluffy during cooking?
6. What nutrients does rice have?
7. Draw and correctly label a diagram of a grain of white rice and of a grain of brown rice. Outline the main nutritional differences between them. Provide reasons for these differences.
8. Outline the types of dishes that use:
   a. long-grain rice
   b. medium-grain rice
   c. short-grain rice
9. List five products made from rice.
10. What is parboiled rice?

CEREALS, BREAD, RICE, PASTA AND NOODLES

## ▶ Let's investigate

1. **a** In small groups, examine a number of recipes that have rice as their main ingredient. Find ten recipes, using a selection of recipe books or visit:
   - www.sunrice.com.au
   - www.betterhealthchannel.vic.gov.au
   - www.yumyum.com
   
   **b** Categorise these recipes as either savoury or sweet. What method is used to cook the rice? What variety of rice is used?
   
   **c** Draw up a table in your workbook like the one below.

| Recipe name | Sweet/savoury | Cooking method used | Variety of rice used |
|---|---|---|---|
|  |  |  |  |
|  |  |  |  |
|  |  |  |  |

2. Create a pamphlet that promotes the importance of eating rice. In order to best present your work, you might like to use a computer design program.

3. **Investigate** why rice products are often consumed by coeliacs. Find out what a coeliac is. The website www.freedomfoods.com.au may be useful.

4. **a** In pairs, research four different savoury rice dishes by using a range of recipe books or websites (refer to question 1).
   
   **b** Undertake a PMI of each recipe—a PMI means you list the 'Pluses', 'Minuses' and 'Interesting points'.

|  | Pluses | Minuses | Interesting points |
|---|---|---|---|
| Recipe 1: |  |  |  |
| Recipe 2: |  |  |  |
| Recipe 3: |  |  |  |
| Recipe 4: |  |  |  |

   **c** Decide which recipe you wish to prepare from your analysis.

5. **Design** your own risotto recipe. Refer to the basic risotto recipe below and add your own flavouring ingredients. Give your original recipe a creative name and then present it to the class. In order to best present your work, you might like to use a computer design program. **Produce** and **evaluate** your risotto.

## Basic risotto recipe

*Ingredients*
½ onion, diced
1 tablespoon butter
½ cup Arborio rice
2 cups stock
30 grams Parmesan cheese, grated

### Method

1. Melt butter in a frypan and sauté onion for two minutes.
2. Stir in rice and cook for two minutes.
3. Stir in flavouring ingredients.
4. Add half a cup of stock and stir until it is absorbed.
5. Add remainder of stock and stir until it is absorbed.
6. Garnish with cheese and serve.

ON THE BOIL

## e-Rice
www.sunrice.com.au

Click on the 'Product' tab and complete the table below in your workbook.

| Rice product | Characteristics | Have you eaten this product? |
|---|---|---|
| Clever rice | | |
| Brown long grain | | |
| Brown medium grain | | |
| Rice cake | | |
| Three-minute rice | | |
| Parboiled rice | | |
| Jasmine rice | | |
| Koshihikari | | |
| Basmati | | |
| Arborio | | |
| Wild blend | | |

## WEB EXTRAS

**www.sakata.com.au**
Sakata produces a diverse range of healthy rice-based snacks.

**www.lundberg.com**
Lundberg Family Farms is a family-owned and -operated farm whose aim is to grow and produce premium brown rice, speciality rice varieties and brown rice products.

**www.worldofrice.com**
Lotus Foods is dedicated to discovering and introducing to the marketplace a range of ancient, rare and new rice varieties.

**www.rga.org.au**
The Ricegrowers' Association of Australia has represented the interests of ricegrowers since 1930. Its aim is to ensure a profitable long-term future for individual ricegrowers and the industry.

**www.riceweb.org**
Riceweb provides information on, among other things, the history of rice, where and how it is grown and recipes.

### www.riceland.com
Riceland is an American site that has some fantastic information about rice.

1. Why is rice considered healthy?
2. Describe how you would correctly store:
   a. cooked rice
   b. uncooked, parboiled and pre-cooked white rice
   c. uncooked brown rice
3. How can the shelf-life of milled and brown rice be extended?
4. Besides water, what other liquids can rice be cooked in?
5. Click on 'Milling'. Describe how these types of rice are produced:
   a. parboiled
   b. brown
   c. white
6. Click on 'Types/forms of rice'. Complete the table below in your workbook.

| Type of rice | Description | Uses in food preparation |
|---|---|---|
| Long grain | | |
| Medium grain | | |
| Short grain | | |

7. Draw a graph to illustrate the different cooking times of:
   a. parboiled rice
   b. brown rice
   c. white rice

CEREALS, BREAD, RICE, PASTA AND NOODLES

## ▶ Puzzled

### Rice match-up

Match the following types of rice with their correct descriptions. Write out your answers in your workbook.

| Rice | Description |
| --- | --- |
| Basmati | This perfumed rice is often served with Thai dishes. The fine, long grains provide a light, fluffy texture when cooked. |
| Jasmine | This plump, round rice is used in risotto as it absorbs the cooking liquid to impart a smooth texture. |
| Calrose | This medium-grain white rice is widely used in Australia. It has a bland flavour. |
| Arborio | This rice is used as an accompaniment to Indian curry dishes as its thin, long grain remains firm and tender when cooked. |

### Scrambled rice

Unscramble the following terms (all contain two words) to discover some products made from rice.

1. ecri kseac
2. erci doonles
3. ceri lfuor
4. kreabstaf realces
5. reic lmki
6. etihw crie

### What rice am I?

What type of rice is used frequently in Australia when preparing Asian, Indian and Mediterranean dishes? Answer the following questions and unscramble the letters in the boxes to work out the answer.

1. What percentage of rice is grown in Asia? _ _ _ ☐ _ _
2. This process is used to explain how rice is sifted to separate the whole grains from the broken grains. _ _ ☐ _ _ _ _
3. This nutrient is found in large amounts in rice. ☐ _ _ _ _ _ _ _ _ _ _ _
4. This Spanish dish has rice and seafood as its main ingredients. _ _ _ _ ☐ _
5. This is a method of cooking rice. _ _ _ ☐ _ _ _ _ _ _
6. When rice is polished, the hard, white _ _ _ _ ☐ _ _ _ _ is left.
7. Wholegrain rice contains more of this nutrient than white rice. _ _ _ ☐ _

Answer: _____

ON THE BOIL

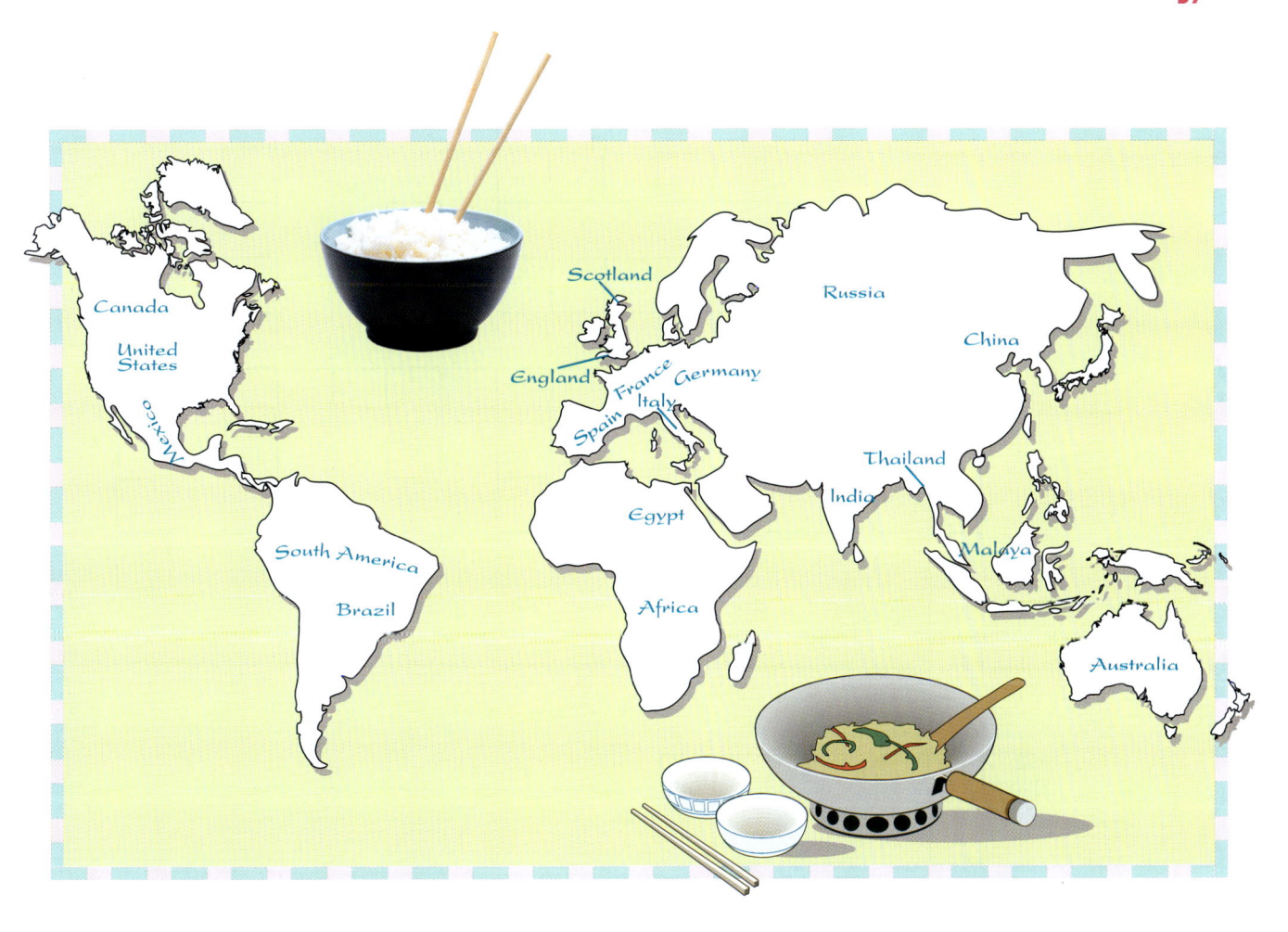

## Around the world

Below are listed many countries whose staple food is rice. Locate them on the map and colour them in:

- Bangladesh
- Brazil
- Burma
- Cambodia
- China
- Colombia
- Dominican Republic
- Gambia
- Guyana
- India
- Ivory Coast
- Japan
- Korea
- Liberia
- Madagascar
- Malaysia
- Mauritius
- Nepal
- Panama
- Philippines
- Singapore
- Sri Lanka
- Surinam
- Thailand
- Vietnam

CEREALS, BREAD, RICE, PASTA AND NOODLES

### Rice crackers

#### Across

1. Word to describe the texture of cooked brown rice
3. Japanese dish that uses koshihikari rice
6. Name of a medium-grain commonly eaten in Australia
9. Method of cooking where rice is placed in cold water
11. The continent that grows a large proportion of the world's rice
12. Process whereby the bran is removed

#### Down

1. Who is said to have influenced the Chinese to eat white rice?
2. What is thrown during weddings as a sign of life and fertility?
4. A Spanish rice dish that uses short-grain rice
5. Name of the Japanese rice god
7. Shape of a rice grain
8. Rice is classified according to the _____ of its grain
10. Rice grain becomes _____ and fluffy when cooked
11. Name of the short-grained rice used in risotto

ON THE BOIL

# Let's Produce

## Rice and chicken wraps (serves 2)

### Ingredients
- 3 teaspoons oil
- ½ brown onion, diced
- 30 grams unsalted peanuts, roughly chopped
- ¼ teaspoon ground cumin
- ¼ teaspoon turmeric
- ⅓ cup medium-grain white rice
- 125 millilitres chicken stock
- 1 teaspoon curry paste
- 100 grams chicken fillets, diced
- 2 flour tortillas
- 4 lettuce leaves, shredded
- 1 tomato, diced
- 2 tablespoons natural yoghurt

### Method
1. Preheat oven to 180°C.
2. Heat one teaspoon of oil in a frypan and add onion. Cook for two minutes.
3. Add peanuts and cook for a further two minutes.
4. Stir in cumin, turmeric and rice.
5. Add chicken stock and mix thoroughly.
6. Bring to the boil and stir over high heat for two minutes.
7. Reduce heat, cover and simmer without stirring until the stock has been absorbed.
8. Remove from heat and set aside for ten minutes.
9. Cover the tortillas with foil and place in the oven for ten minutes or until heated through.
10. Heat the remaining two teaspoons of oil in a frypan.
11. Add the curry paste and stir for one minute, or until aromatic.
12. Add the chicken and cook for about five minutes or until the chicken is cooked.
13. Divide the rice mixture between the warmed tortillas, top with lettuce, tomato, chicken and yoghurt.
14. Roll up firmly to enclose the filling and serve immediately.

## Spicy lamb, vegetable and coconut pilaf (serves 2)

### Ingredients
- ½ cup basmati rice
- 6 teaspoons olive oil
- 2 tablespoons slivered almonds
- ½ brown onion, sliced
- ½ teaspoon cinnamon
- 2-centimetre piece ginger, finely grated
- pinch of cardamom
- ¼ teaspoon ground cloves
- 1 cup vegetable stock
- ¼ cup coconut milk
- 1 carrot, diced and blanched
- 1 zucchini, diced
- ½ teaspoon ground cumin
- ½ teaspoon coriander
- 150 grams lamb

### Method
1. Rinse rice under cold running water until the water runs clear.
2. Heat one teaspoon of oil in a frypan.
3. Toast almonds and remove.
4. Add four teaspoons of oil, onion, cinnamon, ginger, cardamom and cloves and cook for five minutes, stirring occasionally.
5. Stir in the rice, coating it with the oil, and cook for two minutes.
6. Add three-quarters of a cup of stock and bring to the boil.
7. Reduce the heat, cover and simmer for ten minutes.
8. Stir in the remaining one-quarter cup of stock and coconut milk.
9. Add the carrot and zucchini, cover and cook for another five minutes, or until the vegetables are tender.

CEREALS, BREAD, RICE, PASTA AND NOODLES

10 Remove from heat and stir in almonds with a fork to separate the grains of rice.
11 Combine cumin and coriander in a small bowl.
12 Place the lamb on a plate and sprinkle with spice. Rub the spice into the lamb with fingertips.
13 Brush the lamb with one teaspoon of oil and grill until cooked.
14 Diagonally cut lamb into 1-centimetre-thick slices and toss through the pilaf.
15 Serve.

### Cheese and bacon risotto (serves 2)

*Ingredients*
1 tablespoon butter
½ onion, chopped
1 clove garlic
2 rashers bacon, diced
1 cup Arborio rice
1 tomato, diced
3½ cups vegetable stock
2 shakes black pepper
½ cup Parmesan cheese, grated

*Method*
1 Heat butter in a frypan.
2 Add onion and garlic and cook for two minutes.
3 Add bacon and cook for a further two minutes.
4 Add rice and tomato and stir for one minute.
5 Add half a cup of stock to rice mixture and cook, stirring continually until the stock is absorbed.
6 Continue to add half a cup of stock at a time, stirring until it is completely absorbed each time.
7 Remove the frypan from the heat and stir in Parmesan cheese.
8 Season with pepper.
9 Serve and garnish with extra Parmesan cheese.

### QUESTIONS

1 What country does risotto come from?
2 What method was used to cook the rice in this recipe?
3 Describe why Arborio rice is used to make risotto.
4 Besides cheese and bacon, investigate what other ingredients could be used to flavour your risotto.

### Creamy chicken curry (serves 4)

*Ingredients*
4 chicken thigh fillets (or 250 grams chicken fillets, cut into strips)
1 tablespoon oil
½ brown onion
1 tablespoon red (or green) curry paste
200 millilitres coconut milk
2 teaspoons brown sugar
2 teaspoons fish sauce
2 teaspoons lime juice
200 grams potatoes
1⅓ cups jasmine rice
1 tablespoon fresh coriander
lime, for garnishing

*Method*
1 Heat two teaspoons of the oil in a frypan. Add chicken and toss until brown (which takes approximately five minutes). Remove chicken from pan.
2 Add remaining oil to the pan, add sliced onion and curry paste and cook for two minutes.

ON THE BOIL

3. Add coconut milk and gently bring to the boil.
4. Peel potatoes and cut into 2- to 3-centimetre chunks.
5. Add chicken and potato to coconut milk mixture. Simmer uncovered for twenty to twenty-five minutes, or until chicken and potato is cooked.
6. Add brown sugar, fish sauce and lime juice. Serve chicken with steamed rice.
7. Sprinkle with coriander and garnish with a slice of lime.

## QUESTIONS

1. How would you classify jasmine rice?
2. Describe the aroma of the jasmine rice.
3. Describe the texture of cooked jasmine rice.
4. What method did you choose to steam the rice?
5. Describe another way you could garnish your curry.

**e-HINT**
Wasabi is the root of Japanese horseradish. It is very hot, so be careful and eat only a small amount!

### Ingredients
1 cup koshihikari rice, steamed
1 tablespoon rice vinegar
2 sheets nori
2 teaspoons wasabi paste
2 tablespoons Japanese mayonnaise
¼ avocado, sliced
2 Lebanese cucumbers, deseeded
1 carrot, julienne
soy sauce for dipping

## California rolls (serves 2)

### Method

1. Steam rice, add vinegar and allow mixture to cool.
2. Place a sheet of nori on a bamboo mat.
3. Place half a cup of rice on two-thirds of the sheet of nori.
4. Spread some wasabi paste about 2 centimetres from the bottom of the rice; repeat with some mayonnaise.
5. Place a row of avocado, cucumber and carrot.
6. Roll up the nori tightly, ensuring that the first full roll contains all of the ingredients. Repeat steps 2–6 for the other sheet of nori.
7. Wrap in plastic wrap and refrigerate for fifteen minutes. Slice into 2-centimetre slices.
8. Serve with soy sauce and wasabi paste.

## QUESTIONS

1. What is nori?
2. What does the process 'julienne' mean?
3. Why was koshihikari rice used?
4. **Evaluate** the California rolls ☺ ☺ ☹. Did you like them? **Investigate** what other ingredients could be wrapped with the rice? **Design**, **produce** and **evaluate** your own California rolls ☺ ☺ ☹.

**e-HINT**
Wet your hands with water to assist with pushing the rice on to the nori.

# Chapter 7

# Fiddlesticks, Chopsticks: Discover Pasta and Noodles

## e-fact

*Yankee Doodle went to town*
*Riding on a pony*
*Stuck a feather in his cap*
*And called it macaroni*

In eighteenth-century England, the term *macaroni* was used when referring to something exceptional or very good, hence the reference to macaroni in this song written by a British soldier during the American Revolution (1775–1783).

We can trace the origin of noodles back to the Chinese Han Dynasty (206 BC–AD 220). When the island of Sicily was under Arab rule (827–1061 AD), the Arabian people who travelled long distances across the desert are believed to have been the first to make pasta as a means of preserving flour. While many people believe that pasta originated in Italy, explorer Marco Polo is said to have introduced noodles into Italy from China in the late thirteenth century. The Italians then adapted them to create what we now know as pasta.

Pasta and noodle dishes have become popular food choices for many Australians today. This is largely because of the impact of multicultural influences and the increasing number of food products now available.

Pasta and noodles are made from similar ingredients—flour, egg, water and oil—though slightly different methods of preparation and cooking are used. Pasta is most commonly made from flour derived from durum wheat—the hardest of all wheats—as it provides the pasta with firmness when cooked. There are many different types of pasta and they are named according to their shape; however noodles are named according to the type of flour they are made from.

FIDDLESTICKS, CHOPSTICKS

## Classify Pasta and Noodles

**e-fact**
One score equals twenty.

**e-fact**
The diameter of a circle is the distance from one side to the other.

There are literally scores of different kinds of pasta and noodles. We began to eat a lot more pasta dishes after World War II, when many Italians migrated to Australia. With the increase in the number of migrants from Asian countries in more recent years, we have been exposed to an array of noodles, such as Udon and Hokkien, and associated recipes.

**e-fact**
The measurement 3 square centimetres (3 cm$^2$) means a square shape whose sides are each 3 centimetres long.

### Types of pasta

| Type of pasta | Description | Diagram |
|---|---|---|
| Spaghetti | These long, thin, rounded strips of pasta are the most common. Spaghetti is usually associated with the dish spaghetti bolognaise. | |
| Cannelloni | This tubular- or pipe-shaped pasta is approximately 10 centimetres in length and 2–3 centimetres in diameter. | |
| Lasagne | These are thin, flat sheets of pasta. Lasagne sometimes has a ripple edge. | |
| Ravioli | These small squares or parcels are approximately 3 square centimetres and usually filled with meat. | |
| Fettuccini | These long, flat ribbons of pasta are up to 1 centimetre in length and are usually plain or green (spinach is added). | |
| Vegeroni farfalle | This different-coloured pasta is made in the shape of a bow tie. The use of such ingredients as spinach and tomato explains the different colours. | |
| Penne | This tubular-shaped pasta is hollow in the middle, with an angled edge, and usually about 4 centimetres in length. | |
| Macaroni | These small, curved, hollow noodles are approximately 2 centimetres long. They are usually associated with the dish macaroni cheese. | |
| Linguine | These long, flat, thin strips of pasta are similar to spaghetti, but they have square-cut edges rather than round edges. | |
| Risoni | This pasta is the same shape and size as rice, hence the name risoni. It is usually used in soup. | |

CEREALS, BREAD, RICE, PASTA AND NOODLES

## Types of noodles

Although noodles are found to have had origins in China, their popularity has spread to many Asian countries.

Hokkien (or egg) noodles are used in many Asian countries, including Malaysia and Singapore. Made from egg and wheat flour, they are yellow, thick and rounded. Hokkien noodles are sold ready to use as they are pre-cooked and lightly oiled. They are usually used in stir-fries.

Vermicelli noodles are very thin and translucent and made from mung bean starch. They are packaged in blocks and because they are so thin and translucent, they are sometimes referred to as cellophane noodles. Vermicelli can be used in soups and stir-fries.

Rice noodles are made from rice flour and water. Rice paper is also made from rice flour and water and is used to make spring rolls.

Udon noodles are Japanese white wheat-flour noodles that vary in thickness and shape. The Japanese consider the town Shikoku to be the birthplace of this noodle. Each year on 2 July, Udon day is celebrated.

# Properties of Pasta and Noodles

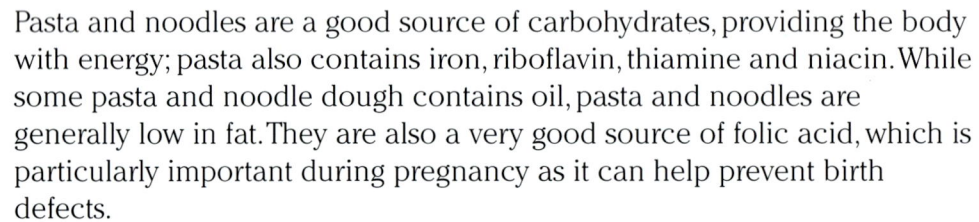

**e-DEFINE**

**Folic acid is a vitamin of the B complex group present in leafy green vegetables, liver and kidney.**

Pasta and noodles are a good source of carbohydrates, providing the body with energy; pasta also contains iron, riboflavin, thiamine and niacin. While some pasta and noodle dough contains oil, pasta and noodles are generally low in fat. They are also a very good source of folic acid, which is particularly important during pregnancy as it can help prevent birth defects.

Pasta and noodles are a good source of protein, containing six of the eight essential amino acids. And by adding meat, you can ensure that you obtain all essential amino acids.

**e-DEFINE**

**Amino acids are organic compounds that occur naturally in plant and animal tissue and that form the components of proteins.**

# Focus on Carbohydrates

Carbohydrates can be classified as either sugars or starches (we often refer to starches as complex carbohydrates). Sugars are the simplest form of carbohydrates and can be more quickly digested than complex carbohydrates, which first need to be broken down into sugars.

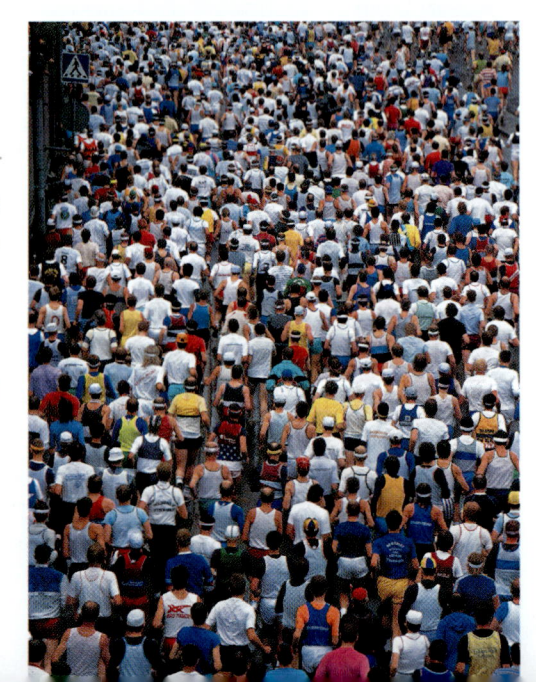

FIDDLESTICKS, CHOPSTICKS

Pasta and noodles contain complex carbohydrates and are often eaten by athletes who participate in long, exhaustive exercises, such as marathons or triathlons. To eat lots of complex carbohydrates prior to these kinds of activity is called carbohydrate loading. By loading up on carbohydrates, the amount of muscle glycogen can increase, assisting with extending energy levels.

## ▶ Let's remember

1. From where did noodles originate?
2. What kind of wheat is used to make pasta? Why is it used?
3. Why have pasta and noodle dishes become popular in Australia?
4. What is the difference between noodles and pasta?
5. Name three kinds of pasta that are hollow in the middle.
6. Which noodles are sometimes referred to as cellophane noodles? Why?
7. What do carbohydrates provide us with?
8. Is pasta a good source of protein? Explain.
9. How can we classify carbohydrates?
10. Why do athletes eat complex carbohydrates?

### e-fact
When cooking pasta, add salt after the water has boiled. This is because water boils more quickly without salt.

## ▶ Let's investigate

1. Choose five different types of pasta and find one suitable recipe for each.
2. Create a collage or picture by gluing different types of pasta onto poster board. Label each type of pasta and be as creative as you like.
3. Create your own limerick about noodles or pasta. Below is an example.

> There once was a young man called Oodle,
> Who loved to eat oodles of noodles,
> He cooked up a storm
> Then kept his food warm
> While he and his girlfriend canoodled!

## ▶ e-Pasta and Noodles

### www.san-remo.com.au

Visit the San Remo website and look at the range of pasta products available. How many different types of pasta and noodles can you count? Go to www.puzzlemaker.com and create your own pasta word search by using the names of the different types of pasta.

### www.heinz.com.au

Visit the Heinz website and answer these questions about canned spaghetti.

1. Heinz Spaghetti is high in _____ and low in _____.

CEREALS, BREAD, RICE, PASTA AND NOODLES

**2** How many tomatoes are in a 425-gram can of Heinz Spaghetti?
**3** What different kinds of children's shapes are available?
**4** Design a new variety of canned spaghetti. What shape would it be and what would you call it? Sketch the shape and label design.
**5** Canned spaghetti in tomato sauce is high in lycopene. Search the Internet and find out what lycopene is.

### www.fantasticsnacks.com.au

**1** What is the origin of Udon noodles?
**2** What are the other names for bean thread noodles?
**3** Name three uses for rice noodles.

### www.asiaathome.com.au

Visit the asia @ home website and find out what other ingredients you need to add to the laksa soup kit (www.asiaathome.com.au/recipes/laksa.htm). Navigate your way around www.colesonline.com.au and do a costing for all of the ingredients required. What is the total cost? Also visit the *Choice* magazine website (www.choice.com.au) and read the information on coconut milk. Write a paragraph on whether or not you think laksa is a healthy food choice.

## ▶ Puzzled

### Pasta box

Form ten words related to pasta and noodles by matching two boxes for each word from the boxes below.

| alle | vermi | nne | cine |
| --- | --- | --- | --- |
| celli | etti | hok | ud |
| agne | on | canne | uine |
| pe | ling | kien | las |
| lloni | fettu | farf | spagh |

### Wordy word

From the word *carbohydrates*, how many other words can you create?

### Pasta mix

Unjumble the letters in the boxes to find the name of a type of pasta.
**1** a nutrient found in pasta and noodles _ _ _ _ _ ☐ _
**2** pasta originated from this country ☐ _ _ _ _
**3** long, thin sheets of pasta _ _ ☐ _ _ _ _
**4** pasta with a meat filling ☐ _ _ _ _ _ _
**5** to cook until just tender _ _ _ _ ☐ _ _
**6** crescent-shaped filled pasta _ _ _ _ ☐ _ _ _

---

**WEBEXTRAS**

www.dalorenzopasta.com.au
Da Lorenzo Pasta is an Australian family-operated company whose product range includes over twenty varieties of fresh egg pasta and a diverse selection of filled pasta, such as cannelloni, pumpkin agnolotti and lasagne.

FIDDLESTICKS, CHOPSTICKS

# Let's Produce

## Ingredients

- ½ cup macaroni
- 1 red apple, diced
- 1 stalk celery, chopped
- 2 tablespoons walnuts, chopped
- 75 grams seedless grapes
- ¼ cup low-fat mayonnaise

### Macaroni Waldorf salad (serves 2)

#### Method

1. Place the macaroni into a saucepan of boiling water. Cook according to the directions on the packet or until al dente (just tender).
2. Drain macaroni and allow to cool.
3. Add apple, celery, walnuts and grapes.
4. Add mayonnaise and stir gently until combined.
5. Serve.

## Ingredients

- 1 rasher bacon, chopped
- ½ onion, chopped
- 1 clove garlic, crushed
- 1 stalk celery
- ½ turnip, chopped
- ¼ carrot, chopped
- ¼ zucchini, chopped
- 1½ cups water
- 1 cup vegetable stock
- ¼ cup macaroni elbow noodles
- ¼ cup cannellini beans, drained
- 1½ tablespoons tomato paste
- 1 tablespoon parsley, chopped
- Parmesan cheese, shaved

### Minestrone soup (serves 2)

#### Method

1. Cook bacon gently.
2. Add onion, garlic, celery, turnip, carrot and zucchini.
3. Add water, stock, macaroni, beans and tomato paste. Cook for thirty minutes.
4. Stir in chopped parsley.
5. Serve, topped with Parmesan cheese.

**e-HINT**
One tablespoon is equivalent to four teaspoons.

## Ingredients

- 1 tablespoon oil
- ¼ onion, sliced
- 100 grams rump steak
- ⅛ red capsicum, sliced
- 150 grams Hokkien noodles
- 1 tablespoon soy sauce
- 1 teaspoon fish sauce
- 1 teaspoon sweet chilli sauce
- 1 stalk bok choy, sliced
- 2 tablespoons cashew nuts

### Beef Hokkien noodles (serves 2)

#### Method

1. Heat oil in a frypan.
2. Add onion and cook for two minutes.
3. Add beef and cook for five minutes, or until browned.
4. Add capsicum, noodles and sauces. Stir.
5. Add bok choy and cook for three to five minutes.
6. Serve immediately, top with cashew nuts.

CEREALS, BREAD, RICE, PASTA AND NOODLES

## Easy cheesy cannelloni (serves 2)

### Ingredients
4 cannelloni shells
100 grams ricotta cheese
100 grams cottage cheese
1 cup silverbeet, shredded
¼ teaspoon nutmeg
¾ cup tomato pasta sauce
¼ cup Parmesan cheese, grated

### Method
1. Mix together ricotta cheese, cottage cheese and nutmeg.
2. Add silverbeet.
3. Place cheese and silverbeet mixture into cannelloni shells.
4. Place in oven-proof dish and top with pasta sauce.
5. Sprinkle with Parmesan cheese.
6. Bake for twenty-five minutes, or until cannelloni is tender.

**e-HINT**
Place a very small amount of pasta sauce on the base of the ovenproof dish. This will stop the cannelloni from sticking to the bottom of the dish!

## Chicken laksa (serves 2)

### Ingredients
1 tablespoon oil
100 grams chicken fillets, sliced thinly
1 shallot, finely chopped
60 grams laksa paste
1 cup chicken stock
1 cup water
250 millilitres coconut milk
½ teaspoon brown sugar
75 grams rice noodles (vermicelli)
¼ cup bean sprouts
1 tablespoon coriander leaves, chopped

### Method
1. Heat oil gently.
2. Add chicken and cook for approximately five to ten minutes until browned.
3. Add shallot and laksa paste and cook gently for one minute.
4. Stir in stock, water, coconut milk and brown sugar.
5. Add vermicelli.
6. Bring to the boil and then simmer for ten minutes.
7. Add bean sprouts and coriander.
8. Serve immediately.

FIDDLESTICKS, CHOPSTICKS

# Cereals, Bread, Rice, Pasta and Noodles: Assessment Task

## WebExtras

- General information on breads and cereals can be found at:
—www.kellogg.com.au
—www.weetbix.com.au
—www.uncletobys.com.au
—www.sanitarium.com
—www.woolworths.com.au
—www.foodwatch.com.au

- Information on *The Australian Guide to Healthy Eating* can be found at:
—www.foodwatch.com.au/shelp_guide.html
—www.health.gov.au/pubhlth/strateg/food/guide/index.htm

This assessment task addresses the outcomes HPIP0501, HPIP0502 and TEMA0501 from the Technology Key Learning Area.

1. **Investigate** your pantry at either school or home and write down all of the food products made from cereal grains.
2. Identify which food products from the list above that you would consume for breakfast, lunch, dinner or snacks. (Some may belong in more than one category.) Complete a table like the one below.

| Breakfast | Lunch | Dinner | Snacks |
|---|---|---|---|
|  |  |  |  |

3. When would you most often eat cereal products? Why?
4. Make a collage to depict a range of cereal products.
5. a. Describe the different types of packaging used for the range of cereal products available. Are they environmentally sound? Can they be recycled? Are they made from recycled products? Outline some ways the packaging could be recycled and/or minimised.
   b. Explain why it is important to minimise waste.
6. Select five different breakfast cereal varieties. Read their labels and complete the table below.

| Name of breakfast cereal | Types of grains |
|---|---|
|  |  |

7. Outline the nutritional benefits of consuming breads and cereals in your diet. Include a discussion of possible dietary-related diseases that may be prevented.
8. **Investigate** the number of servings of cereals recommended for your age group in *The Australian Guide to Healthy Eating*. **Design** a diet that would ensure you meet the recommendations.

Section 3

# Vegetables and Legumes

**The Australian Guide to Healthy Eating** sample serve

75 grams ($\frac{1}{2}$ cup) cooked vegetables
75 grams ($\frac{1}{2}$ cup) cooked dried beans, peas or lentils
1 cup salad vegetables
1 potato

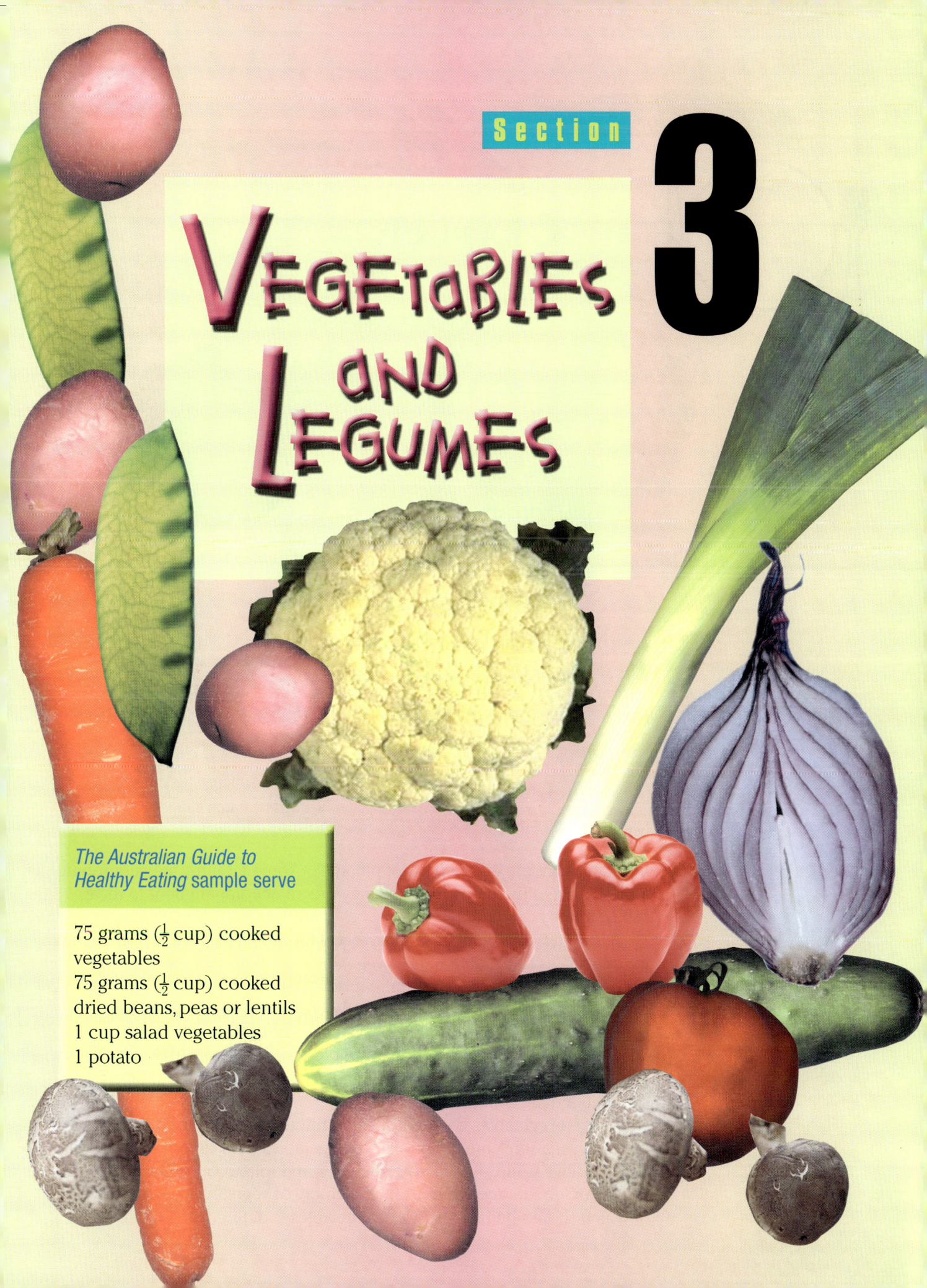

# Chapter 8

# Veg Out: Discover Vegetables

## e-fact

Did you know that the Spanish first invented tomato ketchup? When their explorers brought back some wild tomatoes from Peru, the court chefs used them with oil, vinegar, onions and pepper to make a sauce.

Vegetables are the edible parts of plants. For some families, they contribute a major part of the diet; for others, they are an accompaniment to meat, fish or poultry.

Australia is a very lucky country because many immigrants have introduced us to a great range of vegetables. For example, the Italians and Greeks introduced us to zucchini, capsicum and eggplants.

We should select fresh vegetables that are firm, crisp and not bruised. Vegetables are best purchased in season, when they are most nutritious, flavoursome and cheaper. They should be stored in the crisper section of your refrigerator; however tuber vegetables, such as potatoes, should be stored in a dark, cool place.

Frozen vegetables can be just as nutritious as fresh vegetables—not to mention very convenient. Dehydrated vegetables tend to lose some of their nutritive value, particularly vitamin C and folate.

We are encouraged to consume a variety of vegetables each day, some of which should be raw. A serve of vegetables is defined as 'half a cup of cooked vegetables'. Unfortunately, we are eating fewer vegetables, probably because of the greater number of fast foods available. However supermarkets have a range of ready-to-serve fresh vegetables available to persuade us to eat vegetables more regularly.

## Asian Vegetables—Bok Choy

In Australia the consumption of Asian vegetables has risen: according to recent statistics, there has been a 15 per cent increase in space allocation to oriental foodstuffs in supermarkets over the past two years. During the Gold Rush of the 1850s Chinese migrants brought their traditional vegetables with them when they arrived. Many became market gardeners when the gold ran out; Chinese market gardens are still grown around capital cities today.

VEGETABLES AND LEGUMES

**e-fact**

The Chinese believe that food should be their medicine.

Asian vegetables require quick cooking methods, such as stir-frying and steaming.

One particular Asian vegetable is bok choy. It has white or green stems that are thick and crunchy, and wide green leaves; both the stems and leaves are eaten. Bok choy is an excellent source of vitamin C, provides some iron, fibre and folate and has no fat and minimal kilojoules. It only takes two to three minutes to cook.

# Salad Vegetables

Many supermarkets have a range of ready-to-serve fresh vegetables, meaning they are washed, cut up and packaged. Examples include:
- baby spinach
- carrots
- cauliflower
- broccoli
- lettuce

Selected vegetables are mixed to make such combinations as:
- stir-fry mixed vegetables
- salad mixes
- soup packs

Such vegetable packets encourage us to eat more vegetables because they only take a few minutes to prepare at home.

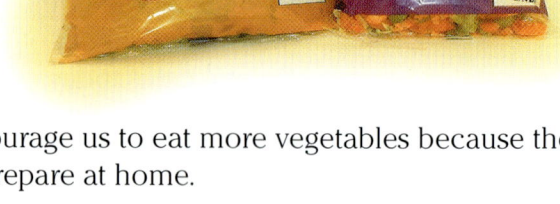

Many types of lettuce are available for us to purchase today. These include:
- iceberg
- rocket
- cos
- radicchio

VEG OUT

# Potatoes

**e-RIDDLE**

Q: When does the Irish potato come from another country?
A: When it is French fried!

**e-RIDDLE**

Q: What do you get when you cross an onion with a potato?
A: A potato with watery eyes!

**e-Fact**

The word *potato* comes from the Indian *pappas*, which became *batata* in Spanish and then potato in English.

Potatoes are Australia's most eaten vegetable. About 70 kilograms is eaten each year; however only 35 kilograms is bought raw, 21 kilograms is bought as French fries and 14 kilograms as potato crisps!

The potato was first grown in South America. Spanish explorers introduced it into Europe in the sixteenth century. English explorer Sir Francis Drake is credited for introducing the potato into England and Sir Walter Raleigh is recognised for introducing it into Ireland.

Potatoes are usually classified according to the texture of their cooked flesh: waxy or mealy. Waxy varieties are more moist and less dense. Their cells tend to remain together when cooked, so they maintain their shape when peeled, sliced and diced. They are ideal for salads and scalloped potatoes. Mealy potatoes are less moist and more dense. Their cells tend to crumble when cooked. They are said to be floury and are ideal for mashing and baking as a light, fluffy texture results.

## Types of potatoes

| Potato | Texture | Uses | Picture |
|---|---|---|---|
| Bintje and pink eye | Waxy | Both are ideal for baking and boiling. | |
| Chats (also known as new potatoes) | Mealy | Chats are great for boiling and roasting. They should be eaten with their skins on. | |
| Desiree | Mealy | Desiree potatoes are one of the best varieties for making mashed potato and gnocchi. | |
| Kipfler | Mealy | Kipfler potatoes are ideal for baking, boiling and mashing. | |
| Pontiac | Waxy | Popular and very versatile, pontiac potatoes are used for baking, mashing and frying. | |
| Purple Congo | Mealy | Purple Congo potatoes make great mash. Their fantastic purple colour adds to their appeal. | |

VEGETABLES AND LEGUMES

# Vegetables and Diet

Vegetables definitely add variety to our diet: they contribute a variety of shapes, sizes, colours, texture and taste!

Studies have shown that populations that eat a higher proportion of vegetables are less likely to develop some cancers. No one particular component in vegetables, or one particular vegetable, is the answer; however it is from consuming a variety of substances in a range of vegetables, such as:
- orange vegetables
- green vegetables
- tomato, capsicum, corn or peas
- potato, sweet potato or yam

We should also ensure that we have at least some cooked vegetables, as well as some raw vegetables.

# Properties of Vegetables

**e-fact**
As vegetables age, they lose their nutritive value.

**e-define**
The Glycaemic Index is a measure of the rate at which carbohydrates are broken down into glucose.

As vegetables come from different parts of plants, each type will vary in its nutrient content. However they generally contain a lot of water, fibre and vitamins and minerals; they are also virtually fat-free.

Vegetables provide us with antioxidants—they have an anti-cancer action within the body. It is sensible to include in the diet a variety of vegetables because their high fibre and water content means they are bulky, thereby satisfying our appetite without contributing excess kilojoules. Vegetables also tend to have a low Glycaemic Index: they release their carbohydrates more gradually into the bloodstream. This is advantageous to people suffering from diabetes, weight problems and heart disease.

The skin of vegetables provides fibre, underneath which are located vitamins and minerals. It is therefore best to either eat the vegetables unpeeled or peel the skin off very thinly. If cooking is required, vegetables should be cooked quickly in minimal water. For example, steaming and microwaving are excellent ways to retain the nutrients.

## Lycopene

Lycopene is a natural red colour in food, such as tomatoes. The redder in colour, the more lycopene the food contains. Lycopene is both a fat-soluble chemical and an antioxidant, which means it is good for our health. Oxidative free radicals are found in our body and are not so good for our health. However lycopene neutralises their negative effects by minimising the damage that these substances can cause to body cells. Lycopene is known to help with decreasing the risk of heart disease and some cancers, such as prostate cancer. The body more readily absorbs lycopene from such processed products as cooked tomatoes and tomato sauces.

# Classify Vegetables

Vegetables can be classified according to the part of the plant that is eaten.

| Part of the plant | Examples of vegetables from this part |
| --- | --- |
| Bulb | Onion, leek, garlic, spring onion |
| Root | Carrot, parsnip, turnip, beetroot |
| Tuber | Potato, sweet potato, yam |
| Seed | Green pea, sweet corn, broad bean |
| Fruit | Tomato, pumpkin, cucumber, capsicum |
| Stem | Celery, asparagus |
| Flower | Cauliflower, broccoli |
| Leaf | Lettuce, cabbage, spinach |
| Fungus | Mushroom, truffle |
| Shoots and sprouts | Mung bean sprouts, alfalfa, snow pea sprouts |

# Focus on Folate

**e-fact**
Folate is also known as folic acid or folacin.

**e-define**
Food that has vitamins added is known as fortified or enriched food.

Folate belongs to the B group of vitamins. It is found in not only various vegetables, such as broccoli, cauliflower and spinach, but also fruit, legumes and nuts. Unfortunately, much of the natural folate in vegetables is lost during cooking, so it is best to consume the vegetables raw. Folate is added to some food, such as breads and cereals.

VEGETABLES AND LEGUMES

Folate is essential for the healthy development of all body cells, the brain, the spinal cord and the skeleton of a foetus. It is often heralded as a miracle cure for birth defects, such as spina bifida, and is believed to protect our bodies from heart disease and colon cancer.

One of the most common problems in industrialised countries is folate deficiency. As a consequence, pregnant women are encouraged to consume folate in their diets.

**Little things make BIG differences**
Victorian Folate Campaign 1999

### ▶ Let's remember

1. Why is it desirable to purchase vegetables when they are in season?
2. Why are we encouraged to consume a variety of vegetables each day?
3. Design a menu for one day that contains at least five serves of vegetables. Make sure it contains foods that you would eat. Remember to include some raw vegetables!
4. Why is there a growing trend for ready-to-serve fresh vegetables in our supermarkets?
5. 'The Chinese believe that food should be their medicine.' In your own words, explain what this statement means.
6. List two reasons why it is best to eat vegetables unpeeled.
7. How should you cook vegetables so that their nutrients are retained?
8. Describe lycopene.
9. What are antioxidants?
10. Raw vegetables are a good source of folate. Why is folate important in our diet?

VEG OUT

# Yard birds

**By CAROLYN HOLBROOK**
Tuesday 25 September 2001

Remember your grandparents' backyard? Chances are there was a Hills Hoist, a chook run, a few fruit trees and a patch, way down the back, dedicated to growing vegetables. The front yard might have been dressed up with roses and azaleas to impress the neighbours, but out the back there was some serious primary production going on.

Now think of your parents' backyard. If they're Anglo-Australian, they probably ripped out the Hills Hoist, chopped down the lemon tree, dug out the vegie patch and covered the lot with lawn. For them, the backyard was a place for barbecuing, drinking beers (or shandies for the sheilas) and basking in the great Australian dream. What need was there to grow your own fruit and vegies when you could buy everything you needed down at the local supermarket?

What a difference a generation makes. Now the wheel is turning and the productive backyard is making a comeback.

Jane Edmanson, Victorian presenter of ABC TV's *Gardening Australia*, is well placed to observe the trend.

'The vegie patch is very definitely back', she says. 'In the last five or six years, young people have really got into growing their own vegetables and herbs. Before that it was mainly older people.'

So what's feeding this renewed hunger for home-grown fruit and vegetables?

There's no doubt it's driven in part by concern for the environment. While their parents cleared the land and took pride in taming nature, younger people are seeing the consequences of an over-reliance on chemical farming.

High school teacher and father of five, Stephen Kennedy, is typical of the new breed of environmentally aware vegie growers. 'With five kids we go through a lot,' he says. 'I try to minimise that with the chooks and the vegies and the fruit trees, so we end up wasting very little really. We try to tread lightly on the earth, which is a nice idea I think.'

Kennedy refuses to use chemicals in his vegetable garden and tries to be philosophical about what he loses to nature.

'The possums and rats discovered the late corn crop and the late tomatoes last year, so we got nothing, which was very disappointing. On our apple tree we got about half and the possums got half, so that worked out fair!'

While environmental and health concerns are playing their part in the resurgence of the vegetable patch, so are our tastebuds. There's a reaction against the insipid produce sold in so many supermarkets these days and an increasing awareness that home-grown vegies taste better.

University of Melbourne academic and keen cook Anthony O'Donnell has been growing vegetables in his Northcote backyard for seven years.

'My initial inspiration came from a cookbook, which made starting a vegetable patch sound really easy,' he says. 'I think if you're interested in food it's just logical to try growing a few things.'

O'Donnell quickly discovered that some vegetables are more rewarding than others. He's noticed little difference between home-grown and bought eggplants and capsicums, but a huge difference with other vegetables.

'You can never buy a tomato that tastes the same as one you grow yourself and you can never buy a zucchini when it's really small and has that subtle, indescribable taste,' he says. 'It's the same with broad beans; you pick them really small and just eat them raw or very lightly cooked.'

But for all the people who take the plunge and plant their own vegetable garden, there are probably twice as many who baulk at the thought of all that effort. So just how much hard work is involved?

'I suppose to initially get a plot up it's hard work if you have to reclaim some land and bring in some soil or compost or turn over lawn,' says O'Donnell. 'Once it gets going there are intense moments in autumn and spring, but most plants will look after themselves if the soil's healthy.'

O'Donnell has managed to maintain his enthusiasm over seven seasons by keeping things simple.

'In summer, I grow tomatoes, zucchini and beans because they're heavy croppers that go for about four months,' he says. 'You only need to concentrate on a few crops that taste good and give you lots of bang for your buck.'

But at the end of the day, it's a process of trial and error.

'It's a bit hit and miss and you learn as you go. Some years are always going to be better than others. I'm not trying to be self-sufficient, so if there's a crop failure, I can always go off to the shop.'

Sure, the supermarket might remain as an ever-reliable backstop, but it's no substitute for the satisfaction of sinking your fingers into the soil and the joy of munching on a simple salad of mignonette and tomato that you've grown yourself.

They may be simple pleasures, but they're ones that people like O'Donnell wouldn't be without.

'Each morning in summer you've got to pick your little zucchini, snip off the flowers from the basil and make sure your tomatoes are tied up. Then you go off to work with your fingers smelling of basil and tomato vine and you're happy for the rest of the day.'

How could a supermarket ever hope to compete with that?

## WHAT'S IN A NAME?

While the concept of growing your own vegetables might be back in vogue, it seems that the term 'vegetable patch' is not. In fact, you're increasingly likely to hear people refer to their vegie patches as 'kitchen gardens' or 'potager gardens'.

'Kitchen garden' traditionally describes a formal vegetable garden attached to a large estate, while potager conveys a more humble French garden, containing a mixture of herbs, vegies and ornamental plants like nasturtiums, daisies and marigolds.

'Gardening has become a very fashionable thing to do in the last ten years and the term vegie patch sounds a bit granddaddy,' says Edmanson. 'But who really cares whether you're growing a potager or a vegie garden or a polystyrene box full of thyme or basil or parsley? It's just a word.'

The trend towards high-density housing means that few people have the luxury these days of hiding a vegetable patch down the back of their quarter-acre block. If you're going to see your vegie garden every time you step out the back door, chances are you want it to look good.

Looks are obviously important to Stephen Kennedy, who's designed the vegetable patch in his Murrumbeena backyard in the shape of a Tudor rose, using wooden garden pickets dug into the ground to create a perimeter of petals!

They're also important to O'Donnell: 'I think my vegie patch looks beautiful at the height of summer. There's zucchini, tomato, basil and maybe a flower plant running wild and beans climbing up the back over the fence. There are bees madly going around pollinating and it's just a summer bounty—I consider it part of my garden and I think the term "kitchen garden" is meant to capture that.'

# VEGETABLES AND LEGUMES

**COMMUNITY GARDENS**

A large part of Australia's urban mythology might be built around the quarter-acre block, but how many people actually live on one these days? If high-density, apartment living has spawned a batch of frustrated vegetable growers, it's also led to the rise of the community garden.

CERES (the Centre for Education and Research in Environmental Strategies) in East Brunswick operates one of the country's longest-running community gardens. Established in 1982, the four-hectare environmental park has about 50 allotments. Members pay a small fee each year and are expected to grow their vegetables organically.

Community gardens are popping up around inner city Melbourne. In 1999, the Port Phillip Council handed over a prime patch of St Kilda land for the establishment of a community garden called Veg Out.

Elwood artist and Veg Out committee member Salvatori Lolicato says the project has had a tremendous response from local residents. 'We have lawyers, doctors, students, pensioners, single mums. There's a real mix.'

The local kindergarten and the St Kilda RSL both have an allotment, as do a handful of well-known actors like *Ballykissangel's* Robert Taylor and *Stingers'* Peter Phelps.

Lolicato's motivation for becoming involved in Veg Out is typical. 'Space is always at a premium in Elwood and St Kilda. I've got a tiny concrete slab of a balcony that doesn't get any sun and I really wanted some open space to grow a few vegies.'

Members also relish the opportunity to socialise and share information. 'There's a wonderful sense of community here. We're all trying to combat the snails and slugs and the birds together.'

Phone CERES on 9387 2609 and Veg Out on 9531 7842. For information about your nearest community garden, phone your local council.

**STARTING OUT**

*Jane Edmanson's tips for the novice vegetable gardener:*

Choose a sunny site that's protected from the elements.

Think small. A space that's 1.5 m x 1.5 m is quite sufficient for starting out.

It's essential that you have fertile soil, so start recycling your organic waste and buy some quality manure.

Begin with easy things like mignonette, cos, silverbeet, spinach, rocket, tomatoes, zucchini and beans. Home-grown corn tastes terrific, but it's a very heavy feeder and low yielder.

Water regularly, which means every morning or second morning in summer.

In the growing season, you'll need to mulch, or cover the soil around your vegies with a carpet of pea straw or old lawn clippings. It prevents water evaporating and stops weeds becoming too prolific.

*The Australian Vegetable Garden* by Clive Blazey from The Digger's Club is a comprehensive guide to growing your own vegies. Phone 5987 1877.

## QUESTIONS

1 Outline three reasons why the vegie patch is making a comeback.
2 According to Anthony O'Donnell, what are the advantages and disadvantages of having a vegie patch?
3 What are the two new trendy names for a vegie patch and identify the differences between them.
4 Why is the appearance of a vegie patch important to people today?
5 What are community gardens?

## Let's investigate

1 List all the vegetables you can think of that you can eat raw.
2 List the many ways you can buy tomatoes. Compare the shelf-life and price of each variety.
3 Investigate the functions of the following nutrients and list several vegetables that contain them:
   a vitamin A
   b vitamin C
   c B group vitamins
4 Calcium and iron are found in some vegetables. Investigate why vegetables are not a good source of these nutrients. (Hint: the body's ability to absorb these nutrients)
5 Visit the supermarket's fruit and vegetable section.
   a Locate and list the number of packaged vegetables.
   b Locate and list the vegetable-mix packages.

VEG OUT

## WEBExtras

**www.fandvforme.com.au**
The Fruit and Vegetables for Me website provides information on fruit and vegetables, as well as games, prizes and links to other sites.

**www.foodwatch.com.au**
Foodwatch provides facts on healthy eating.

**www.betterhealthchannel.vic.gov.au**
Sponsored by the Victorian State Government, the Better Health Channel provides the community with access to online health-related information.

**www.qfvg.org.au**
The Queensland Fruit and Vegetable Growers is an agricultural–political body that represents the interests of the State's fruit and vegetable growers.

6  Research one of the following uncommon vegetables: okra, celeriac, cassava, globe artichoke, kohl rabi, vegetable spaghetti or yam.
  a  Sketch your chosen vegetable.
  b  Find out its colour.
  c  Describe its taste.
  d  From which country did it originate?
  e  How is this vegetable classified?
  f  How is this vegetable used in food preparation? Provide some recipes that incorporate it.
  g  Is there any other additional information you have found?

7  Find out the difference between:
  a  globe and Jerusalem artichoke
  b  red and white sweet potato
  c  broccoli and Chinese broccoli
  d  red and white cabbage
  e  apple and Lebanese cucumber
  f  Belgian and curly endive
  g  Queensland blue pumpkin and butternut pumpkin
  h  green and yellow zucchini

8  Research the following methods to cook vegetables:
  a  stir-frying
  b  steaming
  c  blanching
  d  boiling
  e  microwaving

   **Design** a poster/pamphlet to illustrate what you have found. A computer design program could be used.

9  Record and analyse your own daily vegetable intake. List all of the sources of vegetables that you consume for breakfast, lunch, dinner and snacks.
  a  Did you consume a variety of vegetables each day?
  b  Did you consume raw and cooked vegetables?
  c  Did you consume a variety of vegetables, such as orange and green vegetables, sweet potato, tomato, capsicum, corn or peas?
  d  In what areas, if any, do you need to improve your vegetable intake?
  e  Identify three ways you could improve your vegetable intake.

10 **Investigate** the Irish Potato Famine (1845–1847). Discuss the importance of the potato to the Irish economy.

11 List a vegetable for each letter of the alphabet.

## ▶ *e*-Vegetables

**www.woolworths.com.au**

Go to the Safeway website and find out which vegetables are now in season.

# VEGETABLES AND LEGUMES

## Puzzled

### Match up

Match the words in the first column of the table below with those in the second column to create a vegetable.

| | |
|---|---|
| Apple | Potato |
| Broad | Sprout |
| Jerusalem | Cabbage |
| Snow | Cucumber |
| Pontiac | Lettuce |
| Butternut | Capsicum |
| Spanish | Pea |
| Cos | Pumpkin |
| Sweet | Bean |
| Savoy | Corn |
| Red | Artichoke |
| Brussel | Onion |

### Veging out on vegetables

#### Across

1. I grow under the ground and have two varieties: red and white; I am also known as kumara (two words)
3. My name is derived from the Italian word *brocco*, meaning 'branch or arm'; I am a green vegetable with clusters of green flowers
4. Also known as an aubergine, this vegetable usually has a smooth, shiny, purple skin with a creamy white flesh scattered with lots of brown seeds
6. This is a small, green, round vegetable, sounding like the sixteenth letter of the alphabet
7. This white vegetable is often referred to as the aristocrat of the cabbage family as it has a tight head of flower buds
10. This long, thin vegetable has varieties called broad, butter and snake
11. I am a white-fleshed root vegetable that has a tinge of pink at the top
12. Also called a courgette; I am a traditional ingredient of dishes, such as ratatouille, and am often cooked with tomatoes
14. A green, leafy vegetable that gave Popeye the Sailor muscles
16. I grow on a vine and am popular in the United States on Halloween
17. This is a green, leafy salad vegetable
18. Also known as maize, this vegetable is also a cereal recognised as having

originated in the United States, consisting of round, yellow kernels that form an ear, surrounded by a green husk

## Down

**2** I am a white root vegetable similar in shape to a carrot
**5** An orange root vegetable widely eaten by many people, including Bugs Bunny
**7** This vegetable belongs to the same family as chillies but is milder and sweet tasting; the Americas call it 'bell pepper'
**8** This vegetable belongs to the onion family, and resembles the spring onion but is larger and cylindrical in shape
**9** It is said that the slaves who built the Egyptian pyramids ate this white vegetable, along with garlic and radishes; this vegetable is a bulb and consists of layers that are closely wrapped together, and is capable of making you cry
**13** I am an essential vegetable in coleslaw and sauerkraut
**15** This green vegetable has an edible stem often used in soups and stews, but is most popular as a salad vegetable

VEGETABLES AND LEGUMES

### Vegetable choices

1  Which of the following is not a variety of lettuce:
   **a**  cos          **b**  shuttle          **c**  rocket          **d**  iceberg
2  Which of the following is not a variety of potato:
   **a**  pontiac      **b**  purple Congo     **c**  blue eye        **d**  Bintje
3  Which of the following is not a variety of pea:
   **a**  sugar snap   **b**  garden           **c**  snow            **d**  ginger
4  Which of the following is not a variety of pumpkin:
   **a**  Japanese     **b**  Queensland blue  **c**  butternut       **d**  golden nugget
5  Which of the following is not a variety of tomato:
   **a**  Roma         **b**  cherry           **c**  lima            **d**  common

### Vegetable chop

Match two boxes to make eight vegetables.

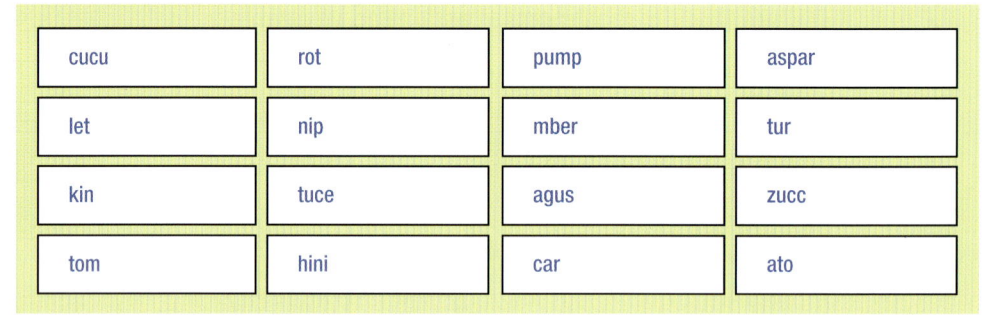

| cucu | rot  | pump | aspar |
| let  | nip  | mber | tur   |
| kin  | tuce | agus | zucc  |
| tom  | hini | car  | ato   |

### Vegetable game

Students are to be organised into groups of four. Their task is to list all of the vegetables they can think of within a five-minute time span. The following classifications can be used to help:
- tubers
- roots
- bulbs
- stems
- shoots and sprouts
- leaves
- flowers
- fruits
- seeds

After five minutes, each group can swap their responses and a scoring system will be completed:
- +2 point for every vegetable listed
- −1 point for every vegetable not listed under their correct classification
- +4 bonus points for every vegetable that was listed by only one group

VEG OUT

# Let's Produce

### Gnocchi  (serves 2)

**Ingredients for gnocchi**
1½ cup mashed potato
1 egg yolk
¼ cup flour

#### Method
1. Mix all of the gnocchi ingredients together to form a firm consistency.
2. Roll into a sausage shape and cut into 3-centimetre lengths.
3. Cook in boiling water until the gnocchi floats to the top.
4. Drain immediately and serve with sauce.

**For tomato sauce**
1 teaspoon olive oil
¼ brown onion, finely chopped
1 clove garlic
200 grams canned tomatoes
2 teaspoons sweet chilli sauce
2 tablespoons chopped fresh parsley
black pepper
20 grams Parmesan cheese
crusty bread

#### Method
1. Heat oil in a frypan and sauté onion and garlic for two minutes.
2. Add tomatoes and chilli and cook for a further two minutes.
3. Stir in parsley and pepper.
4. Add to gnocchi and sprinkle with cheese.
5. Serve with crusty bread.

#### Variations
**Investigate** what other sauces could be used with the gnocchi. **Produce** and **evaluate** your sauce ☺ ☺ ☹.

### Vegetable fritters  (serves 4)

**Ingredients**
200 grams sweet potato, peeled and grated
250 grams potato, peeled and grated
1 clove garlic, crushed
1 tablespoon parsley, chopped
3 shakes black pepper
1 egg, lightly beaten
⅓ cup self-raising flour
3 tablespoons oil
salad vegetables
crusty bread

#### Method
1. Mix sweet potato and potato in a bowl and squeeze out excess liquid.
2. Stir in garlic, parsley, pepper, egg and flour.
3. Heat one tablespoon of oil in a frypan and add heaped spoonfuls of mixture.
4. Cook for about four minutes on each side over low heat.
5. Transfer to a baking tray lined with paper towel and place in a preheated oven.
6. Repeat with remaining vegetable mixture and oil.
7. Serve with salad and crusty bread.

#### Variations
**Design** your own fritters. Alternative vegetables that could be used include parsnip, zucchini, onion and carrot. **Produce** and **evaluate** your own fritters ☺ ☺ ☹.

VEGETABLES AND LEGUMES

### Ingredients
½ cup zucchini, grated
½ cup corn kernels
½ cup broccoli florets
½ cup red capsicum, diced
2 spring onions, sliced
4 eggs, lightly beaten
¼ cup milk
black pepper
3 tablespoons tasty cheese, grated
mixed lettuce
2 tomatoes, quartered

## Vegetable frittata (serves 4)

### Method
1. Grease a large, flat flan dish.
2. Mix the vegetables and sprinkle into the dish.
3. Whisk the eggs, milk and pepper together and pour over the vegetables.
4. Sprinkle with cheese.
5. Cook at 180°C for thirty minutes or until set.
6. Serve with mixed lettuce leaves and tomato.

### Variations
[Design] your own frittata. Alternative vegetables could include leeks and snow peas. [Produce] and [evaluate] your frittata ☺ ☺ ☹.

### Ingredients
1 tablespoon olive oil
½ onion, chopped
1 carrot, diced
1 celery stalk, diced
1 cup Arborio rice
¼ capsicum, diced
2 tablespoons corn kernels
¼ zucchini, diced
½ cup cauliflower florets
½ cup broccoli florets
2 tablespoons Parmesan cheese
3 spring onions, finely sliced
2 cups vegetable stock

## Vegetable risotto (serves 2)

### Method
1. Heat oil in a frypan and sauté onion, carrot and celery for two minutes.
2. Add rice and cook for a further five minutes.
3. Add stock and simmer for ten minutes, stirring occasionally.
4. Add remaining vegetables and cook until liquid is absorbed and ingredients are cooked.
5. Stir through Parmesan cheese and garnish with spring onions.

### Ingredients
2 teaspoons vegetable oil
200 grams lamb, thinly sliced
1 onion, sliced
2 tablespoons water
½ zucchini, sliced thinly and diagonally
75 grams mushroom, sliced
¼ red capsicum, cut into strips
1 baby bok choy, washed and sliced
½ teaspoon freshly grated ginger
1 clove garlic, crushed
1 small red chilli, finely chopped
2 teaspoons soy sauce
1 tablespoon oyster sauce
2 tablespoons freshly chopped coriander
rice

## Bok choy and lamb stir-fry (serves 2)

### Method
1. Heat oil in a frypan and stir-fry lamb for two to three minutes.
2. Remove from pan.
3. Add onion and sauté for one minute.
4. Add water, remaining vegetables, ginger, garlic and chillies.
5. Cover and cook on medium high for five minutes, stirring occasionally.
6. Add lamb, soy sauce and oyster sauce and heat through.
7. Stir in coriander and serve with rice.

# Chapter 9

# Bean There, Done That!: Discover Legumes

Did you know that the term *legume* is the scientific name for the term *pod*? Peas, beans and lentils have pods containing edible seeds. If we eat the seeds of these plants when they are young, they are known as vegetables; however if we eat the seeds when the pods dry up and split open, we call them legumes. Legumes are therefore dried peas, beans and lentils; they are also known as pulses.

VEGETABLES AND LEGUMES

One of the most popular legumes is the soybean (or soya bean). Soybeans have been an important part of the Asian diet for years. They are processed to make:
- tofu (bean curd)
- noodles
- oil
- textured vegetable protein (TVP)
- soy drinks, such as milk
- soy flour
- soy sauce

Lentils are similar to peas and beans: their seeds are red, green or brown. They are one of the oldest foods grown by humans. It is said that the pyramids in Egypt contained legumes that were put there for the pharaohs to consume on their journey into the next world. Obviously, legumes can be kept a very long time!

Many people say they suffer from flatulence when they consume legumes. This is because bacteria do not break down some of the carbohydrates present until they are in the bowel, thereby causing the production of gases. Soaking the legumes and longer cooking will assist with reducing this effect.

Soaking legumes is an important part of the food preparation because it softens them and reduces the cooking time. Cooking improves their digestibility and nutritional value. Legumes can be bought pre-cooked in cans.

Legumes were a staple food for the poor during the Middle Ages in Europe because they were inexpensive and filling. Nowadays, they are used in a variety of dishes, ranging from soups to gourmet meals. When did you last eat legumes?

**e-fact**

An essential amino acid is one that cannot be made by the body. There are twenty-two amino acids, of which eight are essential amino acids.

# Properties of Legumes

Legumes are a highly nutritious food. They tend to have a soft, floury texture and play an important role in the diets of vegetarians—they are a useful alternative to meat.

Most legumes (except for soybeans) are a source of incomplete protein. This means that they do not contain all of the eight essential amino acids that the body needs. Amino acids are the building blocks that make up protein. In order for vegetarians to have a balanced diet, legumes need to be eaten in combination with other food, such as grains or cereals (chickpeas and rice). This combination will provide all of the essential amino acids.

BEAN THERE, DONE THAT!

Legumes tend to be a much cheaper source of protein than meat and tend to have twice as much protein as grains. They are also generally low in fat and provide plenty of fibre. The fibre found in legumes is soluble, helping to lower blood cholesterol levels and control blood sugar levels in diabetics. Legumes are also an excellent source of carbohydrates.

Because legumes come in a variety of colours, they are used to provide a range of appealing colours in dishes.

## Classify Legumes

**e-fact**

Chickpeas are called *gram* in India and *garbanzo* in Spain.

| Peas  | Beans      | Lentils |
|-------|------------|---------|
| Chick | Soy        | Red     |
| Split | Red        | Green   |
|       | Butter     | Brown   |
|       | Lima       |         |
|       | Haricot    |         |
|       | Black-eyed |         |
|       | Cannellini |         |
|       | Borlotti   |         |
|       | Pinot      |         |
|       | Mung       |         |
|       | Lupin      |         |
|       | Kidney     |         |

## Focus on Phytochemicals

Legumes provide us with phytochemicals: natural plant substances. The word *phyto* means plants, so phytochemicals are the chemicals found in plants. Phytochemicals are said to provide protection against diseases, such as cancer, osteoporosis and heart disease. Soybeans have been promoted as being an excellent source of phytochemicals.

# Legume Products

## Soy

> **e-fact**
> Soybean is often referred to as the 'food of angels'.

The soybean is a cream-coloured oval bean. It has been an essential part of the diets of the Chinese, Japanese and Koreans for many centuries. Soy products are also considered beneficial for the prevention of the symptoms of menopause. They provide complete protein—that is, they include the eight essential amino acids normally found only in animal products—and are thus a complete protein. Soybeans are low in saturated fats and free from cholesterol; the polyunsaturated oil is extracted as soybean oil. They also contribute the good omega-3 fatty acids to one's diet. No wonder it is known as super soy!

### Tofu

Tofu is the Japanese word for bean curd. It is made by curdling soy milk. It does not have much flavour—we would describe it as bland. It is pale, creamy white in colour and is highly nutritious. Tofu varies from a soft, silky consistency to a firm and custard-like product. It is used in both sweet and savoury dishes.

### How Tofu Is Made

The soybeans are washed to remove any impurities.

↓

The soybeans are then ground to a paste and cooked in water.

↓

The cooking process extracts the protein from the beans—this protein is in the soy milk. Fibre is extracted from the milk.

↓

The milk is then boiled for a further ten minutes.

↓

A coagulant is added to curdle the milk and turn it into curds and whey. The coagulant may be acid such as lemon juice or calcium sulphate or magnesium chloride.

↓

The curds (solid) are separated from the whey (liquid). The curds are now known as tofu.

↓

The tofu is cut into blocks and packaged, ready for sale.

### Soy milk

Soy milk comes in full- or low-fat varieties, and is suitable for lactose-intolerant people and vegans. It is a good source of calcium and vitamin $B_{12}$. There are even flavoured varieties of soy milk available.

### Textured vegetable protein

Textured vegetable protein (TVP) is made from soy flour, and is minced and soaked in water prior to cooking. TVP readily absorbs the flavour of other foods during the cooking process, and is used as a substitute for minced meat in such dishes as bolognaise sauces, casseroles and burgers.

## ▶ Let's remember

1. What are legumes?
2. Name some products made from soybeans.
3. What does TVP stand for and what is it?
4. Outline the advantages of buying pre-cooked legumes.
5. Why did the poor in the Middle Ages commonly consume legumes?
6. In what ways are legumes used in food preparation?
7. List three nutrients present in legumes.
8. What is an incomplete protein?
9. Consuming legumes is said to bring about flatulence. How can you reduce the chances of this occurring?
10. What are phytochemicals and why are they considered good for us?

**e-DEFINE**

If a person is lactose intolerant, their body cannot produce enough of the enzyme lactase that breaks down lactose (milk sugar) so that it can be absorbed by the body.

## ▶ Let's investigate

1. Using a range of recipe books and the Internet, match the following dishes with their legume ingredient.

| Dish | Legume |
| --- | --- |
| Chilli con carne | Soybeans |
| Dhal | Lentils |
| Hummus | Peanuts |
| Tofu | Navy beans |
| Baked beans | Split peas |
| Split pea soup | Chickpeas |
| Peanut butter | Kidney beans |

2. Collect a variety of legumes, such as kidney beans, baked beans, lima beans and chickpeas. Draw and label them. Note their colour.
3. Conduct some research into the consumption of soy products.
   a. **Design** a questionnaire so that you can interview five people about their consumption of soy products.
   b. Provide them with a list of soy products and ask them if they have ever consumed any of them.

VEGETABLES AND LEGUMES

c Get them to indicate how often they consume them.
d Find out their reasons why they do/do not consume soy products.
e Graph or tabulate your data. To better present your work, you might like to use a computer spreadsheet.
f What conclusions can you draw about the consumption of soy products within your sample group of people?

4 Investigate the importance of soy in the diet.

5 Undertake some research into soy products, such as:
  a tempeh            f soy grits
  b miso              g soy and linseed bread
  c natto             h soy breakfast cereals
  d soy sauce         i soy yoghurt and ice cream
  e soy flour         j soybean oil

6 It is suggested that you have at least 200 milligrams of isoflavone per day.
  a Graph the statistics from the table below.

### e-DEFINE

Isoflavones are a type of phytoestrogen that have a balancing effect on the body's lack of oestrogen during menopause. Legumes, especially soybeans, are high in isoflavones.

| Food | Isoflavone (mg) |
| --- | --- |
| $\frac{1}{2}$ cup canned soybeans | 40 |
| 114 grams tofu (one block) | 38 |
| 250 millilitres soy milk | 20 |
| 114 grams tempeh (one block) | 60 |
| 1 teaspoon miso | 6 |
| $\frac{1}{2}$ teaspoon soy sauce | <1 |
| $\frac{1}{4}$ cup soy flour (defatted) | 44 |
| 1 teaspoon soy grits | 32 |
| 30 grams dried TVP | 94 |
| 2 slices Uncle Tobys soy and linseed bread | 22 |
| 45 grams Soy 'n' Fibre breakfast cereal | 70 |

b Create a diet that would include at least 200 milligrams of isoflavone per day. As a guide, it is recommended that you consume two to three serves of soy per day.

## ▶ e-Legumes

### www.sanitarium.com.au

1 How long have we known that soy products are beneficial to our health?
2 Why are soy products considered nutritious?
3 What are the natural compounds in soy known as?
4 There are many health benefits of consuming soy products.
  a What are these health benefits?
  b Which substances in soy products contribute to these benefits?
5 How many soy products are recommended to be consumed each day for health benefits?
6 List several ways to incorporate soy products into the diet.

BEAN THERE, DONE THAT!

**www.spc.com.au/healthfoods_nutritional.htm**

1. Why are baked beans considered 'healthy'?
2. What type of bean is used to make baked beans?
3. List the nutrients found in baked beans.
4. Outline the amount of fibre found in baked beans compared with that in apples and wholemeal bread.
5. Why are baked beans good for people with high cholesterol?
6. Is flatulence a problem when eating baked beans? Explain.

**WEBExTRAS**

www.sanitarium.com.au
Read the information on Soy Healthy Up and Go and So Good on the Sanitarium website.

## ▶ Puzzled

### Legume multiple-choice questions

Write the stimulus statement and its correct answer in your book.

1. Which of the following is not a legume?
   a lentil  b dried pea  c fresh pea  d dried bean
2. For centuries soybeans have been very important in the diet for which group of people?
   a Asians  b Americans  c Australians  d Britons
3. TVP stands for:
   a total vegetable product  c the vegetable protein
   b textured vegetable protein  d the vegetable product
4. Lentils come in various colours. Which of the following is not a common colour?
   a red  b green  c brown  d purple
5. Legumes are high in:
   a protein and fibre  c fibre and calcium
   b protein and calcium  d protein and sodium
6. Another name for tofu is:
   a lentil  b legume  c pod  d bean curd
7. What is the scientific name for pod?
   a lentil  b legume  c bean curd  d TVP
8. Hummus is made from which legume?
   a chickpeas  b soybeans  c kidney beans  d split peas
9. Phytochemicals are said to assist with the prevention of which diseases?
   a dental caries and heart disease  c heart disease and osteoporosis
   b dental caries and osteoporosis  d diabetes and dental caries
10. Which of the following is not a type of bean?
    a mung  b split  c lima  d kidney

### Scrambled soy

Unscramble the following words to find out the various ways you can purchase soybeans.

1. nncdea
2. ddeir
3. fout
4. osy klim (two words)
5. yso roluf (two words)
6. derttuxe gableetve onetrip (three words)
7. yos caseu (two words)

VEGETABLES AND LEGUMES

### Making soy sense

Complete the following sentences, using the words in the box only once.

| lentils | Japanese | cholesterol | floury | soybean |

1  One of the most popular legumes is the _____.
2  The seeds of _____ are red, green or brown in colour.
3  Legumes contain soluble fibre, which helps to lower blood _____ levels.
4  Legumes are described as having a soft and _____ texture.
5  Tofu is a _____ term for bean curd.

### Soy delicious

Unscramble the tiles to reveal a message.

| HER | U I | CUR | BE | AN | S A | FOR | D. |
| ME | NOT | TOF | NA | | | | |

| TOF | | S A | | | | | |
| | | | D. | | | | |

### Pulse puzzle

Place all of the letters at the bottom in the squares to complete the statement about pulses. Some letters have already been written in for you. Good luck!

|   | L |   | G |   | M |   |   |
|   | R |   | D |   |   | D |
|   |   | E |   | S |   |   |
|   |   | N |   |   | D |
|   |   | N |   | I |   |   |

```
            E       U
    K   A   N   S   I   A   S
    R   E   N   D   R   I   N
A   L   P   G   T   S   L   E   D
B   E   E   E   A   M   E   S   D
```

BEAN THERE, DONE THAT!

## Lots of legumes!

Find each of the words from the box in the puzzle.

- amino acid
- bean curd
- borlotti
- cannellini
- carbohydrate
- chickpea
- dried
- fibre
- flatulence
- haricot
- incomplete
- legume
- lentil
- phytochemicals
- protien
- pulses
- soybean
- split pea
- tofu
- TVP

| D | V | V | A | R | A | G | H | N | T | P | T | D | S | I |
| C | A | R | B | O | H | Y | D | R | A | T | E | L | Y | N |
| C | M | I | U | S | N | B | A | D | F | I | A | H | X | C |
| E | A | Z | T | G | O | E | Y | L | R | C | M | A | O | O |
| M | W | N | F | T | P | Y | A | D | I | P | P | R | D | M |
| U | T | S | N | K | O | T | B | M | D | R | M | I | R | P |
| G | Q | O | C | E | U | L | E | E | O | D | L | C | U | L |
| E | A | I | F | L | L | H | R | T | A | A | T | O | C | E |
| L | H | K | E | U | C | L | E | O | S | N | V | T | N | T |
| C | S | N | I | O | I | I | I | P | B | E | P | E | A | E |
| Y | C | N | T | M | N | F | A | N | T | O | S | T | E | F |
| E | Z | Y | D | I | C | A | O | N | I | M | A | L | B | V |
| O | H | S | P | L | I | T | P | E | A | O | C | H | U | X |
| P | I | H | F | F | I | B | R | E | T | G | Q | G | R | P |
| S | X | Y | N | L | I | T | N | E | L | O | N | Q | Y | Z |

# Let's Produce

## Felafels (serves 2)

### Ingredients
220 grams canned chickpeas, rinsed
$\frac{1}{2}$ onion, chopped
1 tablespoon tahini
$\frac{1}{4}$ cup fresh parsley, chopped
3 shakes black pepper
2 tablespoons wholemeal flour
$\frac{1}{4}$ cup sesame seeds
1 tablespoon wholemeal flour, extra
yoghurt

### Method
1. Place all of the ingredients in a food processor and blend until smooth.
2. Roll spoonfuls of the mixture in extra flour and shape into small balls.
3. Heat two teaspoons of oil in a frypan and cook for five minutes.
4. Serve with yoghurt.

## Chicken wraps and hummus (serves 2)

### Ingredients
1 chicken breast fillet, sliced into 1-centimetre-thick strips
2 tomatoes
$\frac{1}{4}$ Spanish onion
$\frac{1}{4}$ Lebanese cucumber
1 tablespoon fresh mint
2 teaspoons olive oil
2 large pieces of Lebanese bread
100 grams hummus
100 grams yoghurt

### Marinade
$\frac{1}{4}$ cup olive oil
2 teaspoons lemon juice
1 clove garlic
$\frac{1}{4}$ teaspoon thyme
3 shakes black pepper

### Method
1. Make the marinade by combining the oil, lemon juice, garlic, thyme and pepper in a bowl with the chicken. Coat the chicken strips with the marinade, cover and place in a fridge for about thirty minutes.
2. In the meantime dice the tomato, cucumber and Spanish onion and stir in the mint. Set aside.
3. Drain the chicken and heat two teaspoons of oil in a frypan. Stir-fry for about five minutes and drain.
4. Heat Lebanese bread in an oven for about ten minutes.
5. Spread bread with hummus and place chicken strips in a row down the centre.
6. Top with tomato and cucumber salsa and yoghurt and roll up.

BEAN THERE, DONE THAT!

## Vegetable korma (serves 2)

### Ingredients
- 2 teaspoons vegetable oil
- ½ brown onion, sliced
- 2 teaspoons garam masala
- 1 teaspoon turmeric
- 1 clove garlic, crushed
- 1 carrot, peeled and sliced diagonally
- ½ cup vegetable stock
- 150 grams canned chickpeas, rinsed and drained
- 350 grams butternut pumpkin, cut into 2-centimetre dice
- 150 grams green beans
- 200 millilitres coconut milk
- 2 oval pita bread
- natural yoghurt
- mango chutney

### Method
1. Heat oil in a frypan and sauté onion for five minutes.
2. Add garam masala, turmeric and garlic and cook for one minute.
3. Add carrots and stock and bring to the boil.
4. Reduce heat, cover and cook for five minutes.
5. Add beans, chickpeas, pumpkin and coconut milk and bring to the boil.
6. Reduce heat and simmer uncovered for ten minutes.
7. Place the pita bread in a preheated oven for about ten minutes or until heated through.
8. Serve the vegetable korma with the pita bread, yoghurt and chutney.

### Variations

**Design** your own vegetable korma. Alternative vegetables may include leek, sweet potato, potato, snow pea, turnip and corn.

## Vegetable lentil curry (serves 4)

### Ingredients
- 2 teaspoons vegetable oil
- ½ brown onion, sliced
- 1 tablespoon green curry paste
- 100 grams butternut pumpkin, peeled and cut into large pieces
- ½ carrot, peeled and sliced thickly
- 100 grams sweet potato, peeled and cut into large pieces
- ½ cup red lentils, rinsed and drained
- 1 cup vegetable stock
- ½ bunch English spinach, washed
- black pepper
- 2 cups cooked rice
- ¼ cup yoghurt
- 1 tablespoon fresh coriander, chopped

### Method
1. Heat oil in a frypan and sauté onion for two minutes.
2. Stir in curry paste and cook for thirty seconds.
3. Stir in pumpkin, carrot, sweet potato and lentils.
4. Add stock and bring to the boil.
5. Reduce heat and simmer for fifteen to twenty minutes, or until vegetables and lentils are cooked.
6. Stir in spinach and cook for one minute.
7. Add pepper.
8. Serve with rice, topped with yoghurt and coriander.

### Variations

**Design** your alternative curry. Include carrot and potato and use a Balti curry and serve on a bed of couscous. **Evaluate** your curry ☺ ☺ ☹.

VEGETABLES AND LEGUMES

# Vegetables and Legumes: Assessment Task

This assessment task addresses the outcomes HPIP0501, HPIP0502, TEMA0501 and TEMA0502 from the Health and Physical Education and Technology Key Learning Areas. The Sydney Fruit Market's website (www.fandvforme.com.au) can be used as a basis to complete it.

The assessment task is divided into two parts. Select one vegetable for part 1.

## ▶ Part 1 Assignment

Complete the following tasks. Present your information as a poster, data show presentation, webpage, written report, pamphlet or oral report. Make sure you include pictures and diagrams.

1. Provide a description of your chosen vegetable.
2. Identify when this vegetable is available in Australia.
3. Outline the different varieties of this vegetable.
4. What nutrients does this vegetable supply and how does it contribute to a person's overall health? Include a discussion of possible dietary-related diseases it may prevent.
5. Describe a health promotion strategy/campaign that is used to encourage people to consume more vegetables.
6. In what forms is this vegetable sold to the consumer, such as canned, dried and frozen? Include information on how the different types are packaged and labelled.
7. How is this vegetable grown and harvested?
8. Provide a history of this vegetable.
9. Include any interesting facts about this vegetable.
10. Research how this vegetable could be used in composting and why composting is important.

## ▶ Part 2 Practical Activity

Select one recipe that has the chosen vegetable as its main ingredient.

1. **Investigate** what changes you could make to the ingredients in this recipe.
2. **Design** your own original recipe. Write out your new recipe and give it an original name.
3. **Produce** your new vegetable recipe, ensuring you follow hygiene and safety rules in the Food Technology Centre.
4. **Evaluate** your new vegetable recipe ☺ ☺ ☹. Did you like it? What changes, if any, do you need to make?

### WebExtras

**www.olliesworld.com/aus/html/compost.html**
Ollie Recycles provides information on composting.

**www.7aday.coles.com.au**
Coles 7 A Day provides information on the 7-a-Day campaign.

**www.woolworths.com.au/dietinfo/rsa8.asp**
Safeway provides information on labelling.

**www.cannedfood.org/how-food.html#howcans4**
The Canned Food Information Service supports the Australian canned food industry and the suppliers of tin plate and steel cans.

# Section 4

# Fruit Fruit Fruit!

**The Australian Guide to Healthy Eating sample serve**

1 medium piece, e.g. apple, banana, orange, pear
2 small pieces, e.g. apricot, kiwi fruit, plum
1 cup diced pieces or canned fruit
½ cup fruit juice
dried fruit, e.g. 4 dried apricot halves
1½ tablespoons sultanas

# Chapter 10

# Frrrruit!: Discover Fruit

Some fruits, such as grapes, have been a part of civilisation for many thousands of years. So important were grapes that the First Fleet brought grapevines with them and planted them within days of landing on Australian shores. Other fruits have more recent origins, such as

FRRRRUIT!

> **e-fact**
> Banana fossils have been found in Cameroon, West Africa, dating back to 500 BC.

blueberries, which were first grown commercially about 150 years ago in North America; they have only been grown in Australia for the past twenty to thirty years, with their popularity increasing significantly over the past ten years.

We associate fruit with different countries. For example, lychees and mandarins originated from China. Did you know that kiwi fruit originates from China and is also known as a Chinese gooseberry?

Fruit is a fabulous food because it can be eaten as a snack and, unlike many foods, can be eaten at any time of the day—whether it is breakfast, lunch or dinner or even between meals.

## Classify Fruit

Because of Australia's varied climate—tropical climate in the north and a more temperate climate in the south—we are able to grow just about any kind of fruit. The increase in food imports into the country has also made a greater number of fruits available for us to purchase.

We can classify fruit as:

- citrus (orange, lemon, lime, cumquat)—a thick skin, not usually eaten and grown on trees with green waxy leaves
- tropical (mango, pineapple)—grown in the tropics
- stone (apricot, plum)—a large stone in the centre
- berries (strawberry, raspberry)—small, grown on a shrub and usually ending in *berry*
- vine (grape, passionfruit)—grown on a vine
- hard (apple, pear)—firm, with small seeds inside

> **e-riddle**
> Q: Where do baby apes sleep?
> A: In an apricot!

Many fruits can be dried (such as sultanas and prunes), and sometimes they are given a different name after the drying process, like sultana instead of grape and prune instead of plum. However some dried fruits, such as apricot, apple and pineapple, are not given a different name.

FRUIT

## Properties of Fruit

**e-DEFINE**
Rosehips are the edible fruits (hips) of wild roses.

Fruit provides flavour and texture, adds variety to our diet and is a very good source of vitamins (particularly vitamin C), minerals, carbohydrates and water. Generally, it is low in protein and fat.

Fruit contains a substance called pectin, which is a type of insoluble fibre. Pectin can be found in varying amounts in different fruit. For example, apples, berries and currants have a high pectin content. By eating plenty of fruit, we are increasing the fibre content of our diet. More information about fibre can be found in chapter 4, 'Let's Get Cereals'.

## Focus on Vitamin C

Many of us probably first learnt about vitamin C in our studies of Australian history. In the early days of settlement, when people spent many months on ships, conditions were harsh and food was scarce. Consequently, many people began to suffer from a disease called scurvy. We now know that this disease results from a lack of vitamin C, causing the bleeding of gums and bruising. Vitamin C is therefore important to maintain healthy gums, teeth and bones.

Fruit, particularly citrus fruit, is a very good source of vitamin C; tropical fruit, such as mango, paw paw and pineapple, has a high vitamin C content; and blackcurrants and rosehips are the richest source.

Some people believe vitamin C helps to prevent colds and flu, as well as reduce the risk of some types of cancers.

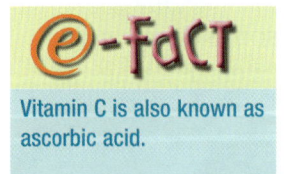

## Fruit Products

**e-fact**
Vitamin C is also known as ascorbic acid.

A major fruit product is jam, which is made by boiling fruit with sugar and forming a gel. Many manufacturers produce jam, including Dick Smith Foods (visit www.dicksmithfoods.com.au), which has a range of jams.

### AS AUSTRALIAN AS YOU CAN GET

The majority of fruit spreads on our supermarket shelves are imported or made by foreign-owned companies. Many contain imported fruit so our farmers suffer. That's why we're fighting back with this premium quality Fruit Spread using Australian grown fruit to ensure the profit and wealth created stays in Australia.

**DICK SMITH'S 5 WAY TEST**
- ✓ Highest quality
- ✓ Proudly Australian made
- ✓ Australian owned
- ✓ Taxes paid here
- ✓ Give our kids a future

A substantial proportion of the profits of Dick Smith Foods go to Australian charities and other important causes.

**DICK SMITH'S Genuine Australian Foods**

ALL NATURAL  
THREE BERRIES  
NO ADDED SUGAR  
100% FRUIT SPREAD  
No Artificial Colours, Flavours or Preservatives

**INGREDIENTS:**
FRUIT 49.5% (STRAWBERRY 36%, BLACKBERRY 9.5%, BLUEBERRY 4%), FRUIT JUICE CONCENTRATE 49.5%, FRUIT PECTIN, LEMON JUICE

**NUTRITION INFORMATION**
Servings size: 15g  Servings per pack: 22

| | Average Quantity per 15g serving | Average Quantity per 100g serving |
|---|---|---|
| ENERGY | 150kJ (35 Cals) | 990kJ (235 Cals) |
| PROTEIN | <1g | <1g |
| FAT | <1g | <1g |
| SATURATED FAT | 0 | 0 |
| CARBOHYDRATE | 8.5g | 57.2g |
| - sugars | 8.5g | 56.7g |
| SODIUM | 5mg | 25mg |

REFRIGERATE AFTER OPENING
MADE FOR DICK SMITH FOODS PTY. LTD.
www.dicksmithfoods.com.au
BY AUSSIE GROWERS FRUITS PTY. LTD.
150-154 MONBULK ROAD, SILVAN VIC 3795
TELEPHONE: 03 9737 9205

330g NET

AUSTRALIAN MADE & OWNED

FRRRRUIT!

## QUESTIONS

Visit the website (www.dicksmithfoods.com.au/dsf5/foods/jams-labels.htm) or look at the label shown to answer these questions.

1 List the ingredients used to make jam.
2 What does the slogan 'We're fighting back' mean?
3 Name three famous Australian brands that have been taken over by foreign countries.

### ▶ Let's remember

1 What kind of fruit crop did the First Fleet bring to Australia?
2 From where do blueberries originate?
3 What is another name for Chinese gooseberry?
4 Why do you think we can grow so many different types of fruit in Australia?
5 What does a plum and a grape become when dried?
6 What is pectin and what is its function?
7 What does a lack of vitamin C cause?
8 What is another name for vitamin C?
9 List the fruit high in vitamin C.
10 What are the two main ingredients in jam?

### ▶ Let's investigate

1 Select ten different kinds of fruit. As a class activity, dissect the fruit and answer these questions.
   a Describe the appearance of the fruit.
   b Describe its texture.
   c Describe its taste. Is it sweet, sour or bitter?
   Cut all of the fruit into bite-sized pieces and make a giant fruit salad for the class to share.
2 Find out what pectin is and investigate its role in making jam.
3 List as many types of fruit as you can that are suitable to be made into jam.

### ▶ e-Fruit

www.freshmarkets.com.au

Search the link to fruit and vegetables information at the Australian Chamber of Fruit and Vegetables Industries website to find as many different kinds of fruit as you can. Then create your own fruit puzzle using www.puzzlemaker.com

www.qfvg.org.au/students/students.html

1 Click on 'Facts' and discover the origin of apricots, custard apples, kiwi fruit, mangoes and pineapples.
2 Draw a map of Australia and then paste or draw pictures of different fruit in the regions where they are grown.

FRUIT

## WEBExtras

**www.chiquita.com.au**
The Chiquita brand's South Pacific Ltd website contains a lot of information about different kinds of fruit.

**www.tropicalfruitworld.com.au**
Visit this site if you are looking for some recipes using some more unusual fruit, such as guava or tamarillo.

**www.sunbeamfoods.com.au**
The Sunbeam Foods website has a wide range of information about dried fruits.

**www.anguspark.com.au**
The Angus Park website has information about dried fruits.

3 Go to the activities section and have some fun with activities of your choice.

### www.bananaland.com.au

1 How many different varieties of bananas can you find?
2 Write a couple of paragraphs to explain why bananas are referred to as brain food.
3 Besides Coffs Harbour, where else are bananas grown?

### www.woolworths.com.au/recipes/listfamiliesingroup.asp?groupid=5

The Safeway website contains a large list of fruit, with an explanation of each. Write down a description of the following more unusual types of fruit.

1 abiu
2 babaco
3 carambola
4 fejoia
5 guava
6 jack fruit
7 nashi
8 pepino
9 pomelo
10 rambutan
11 sapodilla
12 soursop

### www.orchy.com.au

1 What is the difference between orange fruit juice and orange fruit drink?
2 What type of packaging is used for the long-life juice? Describe this type of packaging.
3 Explain the steps involved in processing oranges to produce orange juice.

## ▶ Puzzled

### Fruit and colour

Sometimes fruit has the same name as a colour. Besides orange and lemon, how many can you think of that are also a colour?

### Which fruit?

Name the fruit with the following varieties:
1 Jonathan
2 Bartlett
3 lady fingers
4 Valencia
5 sultana
6 morello
7 clingstone
8 pink

### Missing fruits

Complete these sentences with the correct fruit.
1 In some parts of Australia, cantaloupe is known as _____.
2 The country of origin for kiwi fruit is _____. This fruit is also called _____.
3 When a plum is dried, it becomes a _____.
4 At the Wimbledon Tennis Championships, it is traditional to serve _____ and cream.
5 The biggest _____ in the world can be found at Coffs Harbour.
6 English musician Sir Bob Geldof has a daughter called _____.
7 The cross between a mandarin and a tangerine is a _____.
8 _____ is an anagram of mile.
9 _____ are used to make wine.

# Fruit criss-cross

## Across

3. When this fruit is dried, it is known as a prune
4. This fruit rhymes with tango
7. This fruit is one of the richest sources of vitamin C
10. This fruit is the main ingredient in a dessert named after Dame Nellie Melba
11. We cook and eat the stalks of this fruit; however the leaves are poisonous
12. Varieties of this fruit include Packham and Barlett
14. We often use the pulp of this fruit on Pavlova
15. This fruit is used to make wine
16. Small, red stone fruit that is traditionally available around Christmas time
17. This yellow tropical fruit can also be found in pyjamas

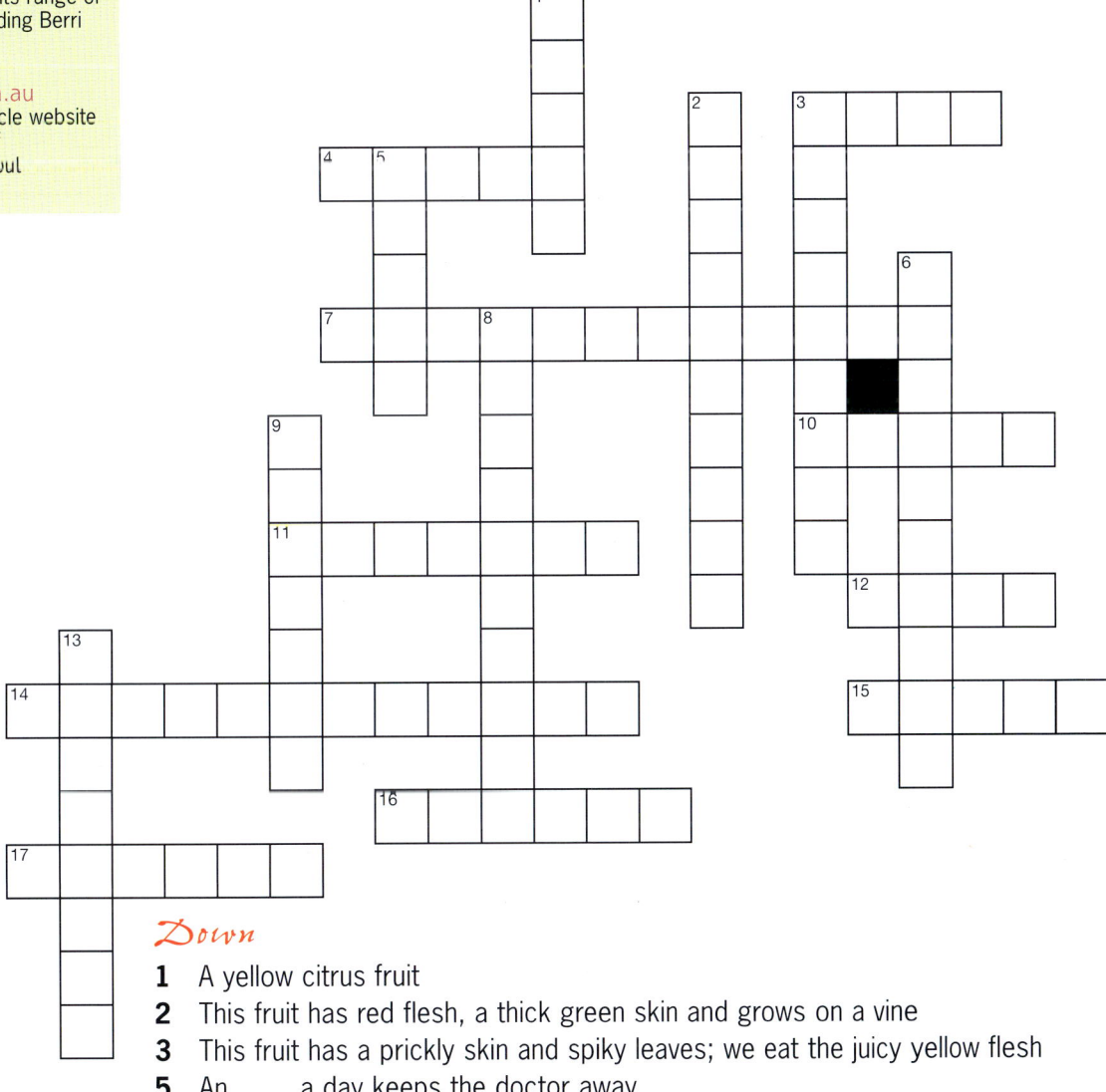

## Down

1. A yellow citrus fruit
2. This fruit has red flesh, a thick green skin and grows on a vine
3. This fruit has a prickly skin and spiky leaves; we eat the juicy yellow flesh
5. An ____ a day keeps the doctor away
6. A type of berry that is commonly served with cream
8. A type of melon that is also known as a rockmelon
9. An orange stone fruit
13. A small orange citrus fruit with a skin that is very easy to peel

## WebExtras

**www.spc.com.au**
The SPC website provides information on its canned fruit products.

**www.cottees.com.au**
The Cottees website contains information on such products as cordial and jam.

**www.berriltd.com.au**
The Berri website provides information on its range of products, including Berri juices.

**www.gcl.com.au**
The Golden Circle website features lots of information about pineapples.

FRUIT

# Let's Produce

## Banana smoothie (serves 1)

**Ingredients**
1 cup milk
1 small banana
1 teaspoon honey
½ teaspoon vanilla essence

### Method
1. Chop the banana roughly.
2. Mix together all of the ingredients and place in a food processor until smooth.

**e-HINT**
If the honey is thick and hard to get out of the jar, place it in the microwave for twenty to thirty seconds.

## Easy individual pavlova (serves 1)

**Ingredients**
1 egg (large) white
½ cup caster sugar
¼ teaspoon vanilla
½ teaspoon cornflour
½ teaspoon white vinegar
2 tablespoons boiling water
cream
selection of fruit

### Method
1. Make sure egg white has no traces of egg yolk.
2. Place all of the ingredients, except cream and fruit, into a bowl and mix with an electric beater until mixture is very firm and peaks form.
3. Spread into a round shape on a tray lined with baking paper.
4. Place in a moderate oven (180°C) for ten minutes.
5. Reduce heat to 150°C for approximately thirty minutes.
6. Cool in oven.
7. Serve with cream and fruit, such as passionfruit and berries of your own choice.

## Apple and rhubarb crumble (serves 2)

**Ingredients**
2 stalks rhubarb, chopped
1 apple, peeled and sliced thinly
1 tablespoon sugar
1 tablespoon water
20 grams butter
¼ cup self-raising flour
1 tablespoon coconut
1 tablespoon brown sugar

### Method
1. Place rhubarb, apple, sugar and water into a saucepan and cook gently on low heat until the fruit is soft and cooked through.
2. Place the fruit into two greased ramekin dishes.
3. Rub butter into flour using fingertips.
4. Add coconut and brown sugar.
5. Place crumble mixture on top of fruit.
6. Bake in a moderate oven for fifteen to twenty minutes, or until the topping is brown.

FRRRRUIT!

## Chicken and pear (serves 1)

**Ingredients**
1 chicken fillet (approximately 150 grams)
30 grams butter
1/4 cup brown sugar
2 tablespoons water
1/2 pear, cut into four pieces

### Method
1. Slice chicken into strips.
2. Melt butter in an oven-proof dish.
3. Add brown sugar and water.
4. Place chicken and pear into the dish.
5. Cover tightly and bake for thirty minutes, or until chicken is cooked and pear caramelised.

## Peach pillow (serves 1)

**Ingredients**
1/2 canned peach
1/4 sheet ready-rolled puff pastry
1 egg white, lightly whisked
1/2 teaspoon caster sugar
ice cream

### Method
1. Place peach, round side up, onto a square of pastry.
2. Cut pastry into a round shape, leaving a 1-centimetre seam around the peach.
3. Brush the pastry with the egg white and sprinkle with caster sugar.
4. Bake at 220°C for ten minutes, or until pastry is golden brown.
5. Serve with ice cream.

# Fruit: Assessment Task

This assessment task addresses the outcomes HPIP0501, HPIP0502, TEMA0501 and TEMA0502 from the Health and Physical Education and Technology Key Learning Areas. The Fruit and Vegetables for Me website (www.fandvforme.com.au) can used as a basis to complete it.

The assessment task is divided into two parts. Select one fruit for part 1.

## ▶ Part 1 Assignment

Complete the following tasks. Present your information as a poster, data show presentation, webpage, written report, pamphlet or oral report. Make sure you include pictures and diagrams.

1. Provide a description of your chosen fruit.
2. Identify when this fruit is available in Australia.
3. Outline the different varieties of this fruit.
4. What nutrients does this fruit supply and how does it contribute to a person's overall health? Include a discussion of possible dietary-related diseases it may prevent.
5. Describe a health promotion strategy/campaign that is used to encourage people to consume more fruit.
6. In what forms is this fruit sold to the consumer, such as canned, dried and frozen? Include information on how the different types are packaged and labelled.
7. How is this fruit grown and harvested?
8. Provide a history of this fruit.
9. Include any interesting facts about this fruit.
10. Research how this fruit could be used in composting and why composting is important.

## ▶ Part 2 Practical Activity

Select one recipe that has the chosen fruit as its main ingredient.

1. **Investigate** what changes you could make to the ingredients in this recipe.
2. **Design** your own original recipe. Write out your new recipe and give it an original name.
3. **Produce** your new fruit recipe, ensuring you follow hygiene and safety rules in the Food Technology Centre.
4. **Evaluate** your new fruit recipe ☺ ☺ ☹. Did you like it? What changes, if any, do you need to make?

---

### WEBExtras

**www.olliesworld.com/aus/html/compost.html**
Ollie Recycles provides information on composting.

**www.7aday.coles.com.au**
Coles 7 A Day provides information on the 7-a-Day campaign.

**www.woolworths.com.au/dietinfo/rsa8.asp**
Safeway provides information on labelling.

**www.cannedfood.org/how-food.html#howcans4**
The Canned Food Information Service supports the Australian canned food industry and the suppliers of tin plate and steel cans.

Section 5

# Milk, Yoghurt and Cheese

**The Australian Guide to Healthy Eating sample serve**

250 millilitres (1 cup) fresh, long-life or reconstituted milk
½ cup evaporated milk
40 grams (2 slices) cheese
200 grams (1 small carton) yoghurt
250 millilitres (1 cup) custard

# Chapter 11

# Milk It!: Discover Milk

Milk is the first food you consumed as a baby, and it is an important food for people of all ages. Most of the milk we drink comes from cows; however we can also drink the milk of goats and even sheep. Breast milk is recommended for babies less than twelve months old.

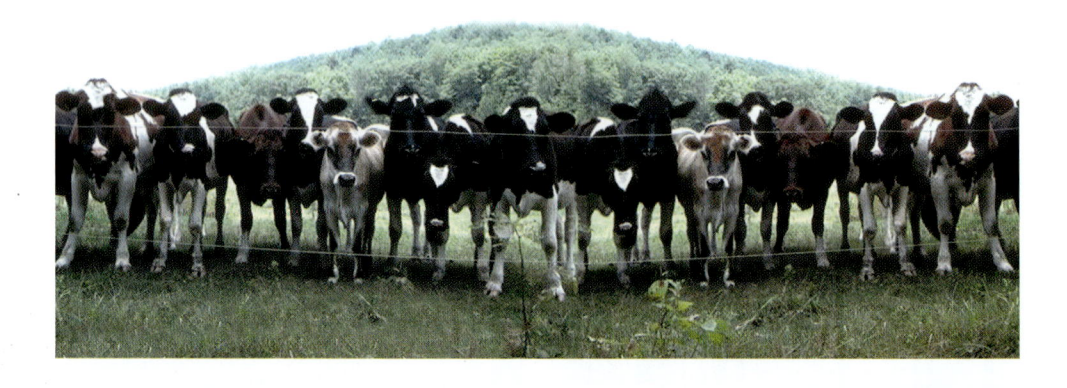

MILK IT!

Once the cows produce milk, it is put into large tankers and taken to a dairy for processing, which includes pasteurisation and homogenisation. During pasteurisation, the milk is heated and the harmful bacteria are killed; its taste and nutritional value are not affected. Homogenisation is when the milk is shaken in a centrifuge so that the fat particles are distributed evenly.

## How Much Do I Need Every Day?

According to *The Australian Guide to Healthy Eating*, people over four years should eat two to five serves of milk, yoghurt or cheese each day. This can include:
- 250 millilitres of milk
- half a cup of evaporated milk
- 40 grams (2 slices) of cheese
- 200 grams of yoghurt
- 250 millilitres of custard

## Classify Milk

**e-DEFINE**

UHT milk is milk that has been sterilised by heating to a very high temperature for a short period of time to kill all of the bacteria.

There are many different types of fresh milk, such as no fat, reduced fat, full cream, calcium enriched and flavoured.

Most people consume fresh milk; however many other forms are available, such as:
- Evaporated milk has had approximately 60 per cent of the water removed. It is stored usually in cans.
- Sweetened condensed milk is the same as evaporated milk, with the exception of the addition of sugar. It is stored in cans or tubes.
- Dried or powdered milk has had almost all of the water removed, and is available as full cream or skim milk (where all of the fat is removed). It is available in cans or packets.
- Ultra-high temperature (UHT) milk is available in cartons and can be kept in the pantry until opened, when it must be refrigerated.

MILK, YOGHURT AND CHEESE

## Properties of Milk

Milk provides us with protein, which is needed for the growth and repair of body tissues. It is also a very good source of calcium and provides fat, carbohydrate and some B group vitamins, such as vitamin $B_{12}$.

## Focus on Calcium

Milk is a very valuable source of calcium because it is needed for healthy bones and teeth. Until the time you stop growing , you have the opportunity to reach your peak bone mass (PBM). This means that the more calcium-rich foods you eat, the more chance you have of depositing calcium in your bones.

As you grow older, you will lose more calcium from your bones than you can deposit, so it is really important to have a good start. Making sure you have lots of milk, cheese and yoghurt can help with this. It is also vital to exercise: exercise and eating foods high in calcium can help prevent osteoporosis later in life.

Osteoporosis is a condition where the bones become brittle and can break; it can affect both women and men. Do you know any older people who have had a simple fall, yet broken a bone? This is possibly because they have osteoporosis. Make sure you do not get osteoporosis! Eat foods high in calcium and participate in regular physical activity.

# Milk Products

Milk products include not only the many different types of milk but also all of the products made from it, such as yoghurt, butter, cream and ice cream. While butter and cream are milk products, it is important to remember that they mainly consist of fat and so should be eaten in small amounts.

## ▶ Let's remember

1. What does pasteurisation mean?
2. What happens in a centrifuge?
3. What is the difference between evaporated and condensed milk?
4. Explain what UHT milk is.
5. Which nutrients are found in milk?
6. What does PBM stand for?
7. Describe the condition osteoporosis.
8. Who is most at risk of suffering from osteoporosis?
9. How can osteoporosis be avoided?
10. List four products that can be made from milk.

MILK, YOGHURT AND CHEESE

# CHILDHOOD RECOLLECTIONS

I remember when I was a little girl I would wake in the morning to the sound of the draft horse clip-clopping down the street outside. Early each morning, the milkman (we called him the Milko) would run from house to house delivering the milk while the horse just kept a steady pace, pulling the milk cart along behind. Somehow the horse knew which houses to stop at as the Milko kept running backwards and forwards from the milk cart to the houses.

Every night we used to leave a milk carrier at our front gate with three empty, clean glass bottles and the exact money for three fresh bottles of milk. Every morning, my brothers and I would fight over who got to pour the creamy layer at the top of each bottle onto our cereal. If there was ever a fight between us, Mum would shake the bottle so that the creamy layer and the milky bottom were all mixed up and no one got the creamy part.

I remember that if we didn't go outside and collect the milk straight away, sometimes it would be warm from sitting out in the sun, just like it used to be at school by the time morning playtime arrived. I hear that the milk program in primary schools has returned and that these days they put the milk in a refrigerator instead of delivering it in a crate and leaving it outside the classroom door. We weren't allowed out to play until we all had drunk our carton of milk. I wonder if it is still the same today?

## QUESTIONS

1. Ask your parents and grandparents about their childhood memories of milk.
2. Why do you think the milk program has been reintroduced into primary schools?
3. Which process mixes the fat particles evenly throughout the milk?
4. What happens to milk if it is left out in the sun?
5. Milk used to be sold in bottles. Investigate the different ways milk is sold now.

## e-Milk

**www.healthybones.com.au**

The Healthy Bones website is a joint initiative of Osteoporosis Australia and the Australian Dairy Corporation aimed at creating an awareness of the impact of osteoporosis on the lives of Australians. Think of an activity that you can do to promote Healthy Bones week at your school. As a reward, ask your teacher if you can access the class activities at www.healthybones.com.au/Fun/default.htm They are loads of fun!

**www.pauls.com.au/products/products.cfm**

Pauls Limited is one of the major manufacturers of milk in Australia. They produce a range of milks, yoghurts and dairy products. Check out their product range for

### WEB EXTRAS

www.dairycorp.com.au
The Australian Dairy Corporation's major objective is to enhance the profitable production and marketing of Australian dairy products.

MILK IT!

the different types of milk sold in Queensland. Look at the advertising for these different types and identify who you think they would appeal to. For example, Cream Top is 'for those who like milk the way it used to be'. Think back to the case study. Who do you think is the target market for the Cream Top product?

### www.puramilk.com.au/great_brands

Pura Milk market many varieties of milk, such as Pura Whole Milk, Pura Light Start, Pura Tone, Pura Edge, Pura Gold and Pura Café.

1  Refer to the Pura Milk website to complete the table below.

| Milk type | Kilojoules (kj) | Protein | Fat | Calcium | Iron |
|---|---|---|---|---|---|
| Whole Milk | | | | | |
| Light Start | | | | | |
| Tone | | | | | |
| Edge | | | | | |
| Gold | | | | | |
| Café | | | | | |

Nutrient content per 100 millilitres of milk

2  Write a few paragraphs comparing the different types of milk according to nutrient and energy (kj) content. Include an explanation as to why the different types of milk have been produced. Which sections of the population are they aimed at?

## ▶ Puzzled

### Milk jingle

Write your own jingle or song about milk using these words: milk, calcium, osteoporosis, milk shake, ice cream, yoghurt and bones.

### Milk match

Complete these statements by using the words from the box below.
Each word can be used only once.

|  |  |  |  |  |  |
|---|---|---|---|---|---|
| yoghurt | fat | sugar | protein | bones | cheese |

1  Flavoured milk contains more _____ than regular milk.
2  Calcium is required for healthy _____ and teeth.
3  Cream and butter are milk products, which are high in _____.
4  The nutrient _____ is needed for the growth and repair of body tissues.
5  If I do not have milk each day, I can always eat _____ or _____ instead.

### Milk scramble

Unscramble these words related to milk

1  ducreed
2  aft owl (two words)
3  kmis
4  paaeevortd
5  ncddsenoe

### WEBExtras

www.nestle.com.au/schoolprojects/milk
This Nestlé milk webpage features a memory game. By completing questions 1 to 12, you learn about how milk is preserved.

www.tasdairy.com.au
The Tasmanian Dairy Industry Authority website not only provides information on health and nutrition but also has project information on the dairy and cow pages and lots of fun activities and competitions.

www.foodsci.uoguelph.ca/dairyedu/fluid.html
The University of Guelph webpage provides an overview of the range of dairy products from milk that are available.

MILK, YOGHURT AND CHEESE

# Let's Produce

## Chocolate cherry dessert (serves 2)

### Ingredients
⅓ cup canned pitted cherries, drained
⅓ cup self-raising flour
1 tablespoon cocoa
⅓ cup milk
¼ cup caster sugar
15 grams butter
½ egg
½ teaspoon vanilla essence
icing sugar

### Method
1. Grease two one-cup ramekins with butter.
2. Place cherries in the base of each ramekin dish.
3. Place flour, cocoa, milk, caster sugar, softened butter, egg and vanilla essence in a food processor and mix until smooth.
4. Pour the mixture evenly over the cherries.
5. Bake in a moderate oven for thirty minutes until golden brown.
6. Serve immediately, dusting with icing sugar.

## Irish soda scones (serves 1)

### Ingredients
1½ cups plain flour
½ teaspoon cream of tartar
¾ cup buttermilk
½ teaspoon salt
½ teaspoon baking powder
jam and cream

### Method
1. Sift dry ingredients together.
2. Make a hollow in the centre and add enough buttermilk to make a soft dough.
3. Turn the dough onto a floured board and knead quickly and lightly until it is free from cracks.
4. Roll out until it is 0.5 centimetre thick and cut into rounds.
5. Place onto an oven tray lined with baking paper and cook at 200°C for ten to fifteen minutes.
6. Serve warm with jam and cream.

### QUESTIONS
1. Which ingredient/s cause the scones to rise?
2. What would happen to the dough if it was overkneaded?
3. What is buttermilk?

## Penne and chicken bake (serves 4)

### Ingredients
25 grams butter
½ onion, chopped finely
2 tablespoons flour
500 millilitres milk
2 tablespoons seeded mustard
100 grams cooked chicken, chopped
150 grams penne pasta, cooked
1 cup broccoli, blanched
125 grams tasty cheese, grated
60 grams Romano cheese, grated

### Method
1. Grease casserole dish.
2. Melt butter in a saucepan. Add onion and cook gently until soft.

MILK IT!

3  Add flour and cook for a further two minutes.
4  Add milk and stir until mixture boils and sauce thickens.
5  Stir in mustard, chicken, pasta, broccoli and tasty cheese.
6  Top with grated Romano cheese.
7  Bake in a moderate oven for fifteen to twenty minutes, or until mixture is heated through and brown on top.

## QUESTIONS

1  What do the terms *blanch* and *roux* mean?
2  **Investigate** what kind of cheese Romano is.
3  **Investigate** what other ingredients you could replace the chicken and broccoli with.

## Passionfruit ice cream (serves 6)

### Ingredients
2 teaspoons cornflour
2 tablespoons sugar
250 millilitres cream
250 millilitres milk
175 grams passionfruit pulp

### Method
1  Mix together cornflour, sugar and cream in a saucepan.
2  Add milk and heat, stirring until mixture boils and thickens.
3  Remove from heat and cool slightly.
4  Pour into a plastic bowl and cover with plastic wrap.
5  Place in the freezer for two hours, or until the mixture starts to set.
6  Place the mixture in an electric mixer on medium speed for about four minutes. Add passionfruit pulp. Stir well.
7  Put into a container and place back into the freezer until firm and ready to serve.

### Variations
1  Replace passionfruit pulp with mango pulp.
2  Use frozen raspberries, strawberries, blackberries or blueberries instead.
3  For an extra-rich ice cream, use double cream instead.

## Peach trifle (serves 2)

### Ingredients
2 slices commercial Swiss roll
2 tablespoons orange juice
2 canned peach halves
2 teaspoons sugar
2 teaspoons cornflour
1 egg yolk
½ cup milk
¼ teaspoon vanilla

### Method
1  Soak the Swiss roll in orange juice.
2  Place the Swiss roll in a serving dish and top with peach half.
3  Blend cornflour with a little of the milk to make a smooth paste.
4  Add the remaining milk, sugar and lightly beaten egg yolk.
5  Heat gently, stirring constantly until the mixture thickens.
6  Pour over peach and chill.
7  Serve.

# Chapter 12

# Need a Little Culture?: Discover Yoghurt

> **e-fact**
> According to legend, an angel whispered the secret of souring milk to Abraham, the first patriarch of Israel who is spoken of in the Bible's Book of Genesis. Abraham is said to owe his masculinity and long life to his consumption of yoghurt.

Yoghurt is a dairy product made from sour milk. Before the introduction of refrigeration, milk was preserved by making it into yoghurt.

People in eastern Europe and western Asia have been making yoghurt for thousands of years. They believe it enhances their life expectancy.

Yoghurt became more popular in Australia after World War II, when many immigrants served it in cafés and restaurants.

To make yoghurt, the milk is first treated to kill the bacteria present and then injected with a bacteria starter and kept warm. The bacteria in the starter could be acidophilus and bifidobacterium, which are simply known as A and B cultures; both are considered to be 'good' bacteria because they can assist with maintaining healthy intestinal functioning. The bacteria begin to grow quickly in the milk, producing acids that break down lactose. Lactic acid is then released, causing the milk to become sour.

The souring of milk results in the protein clumping together and causing the milk to thicken (or coagulate) into a firm curd. The lactose

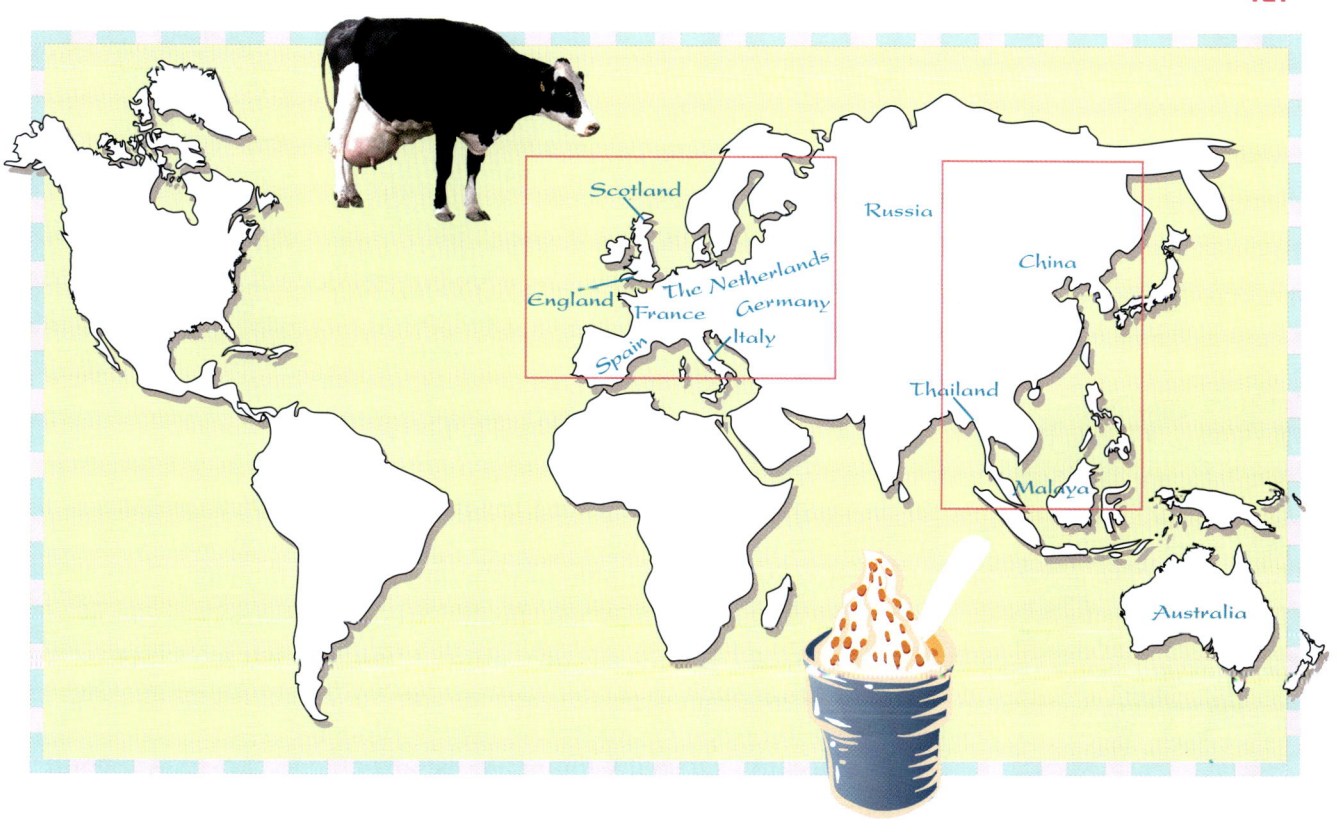

NEED A LITTLE CULTURE?

121

### e-fact
The word *yoghurt* comes from the Turkish *jugurt*, meaning soured milk.

### e-define
Lactose is the name of the sugar found in milk.

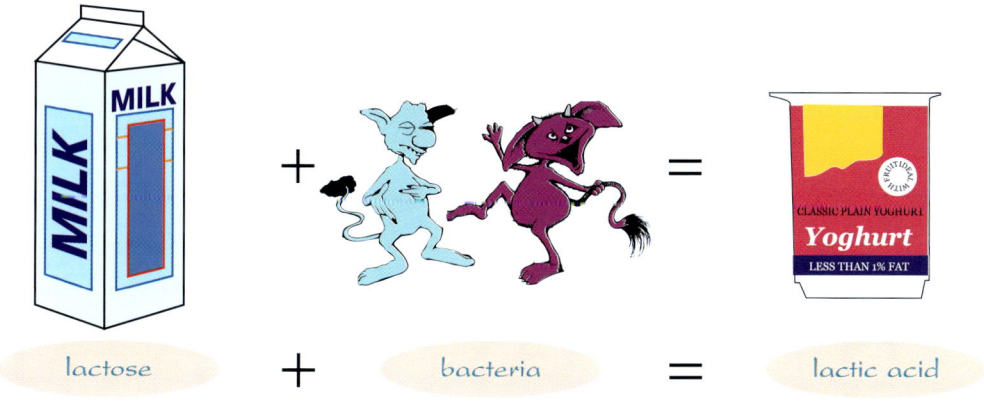

### e-fact
The slightly acidic taste or 'tang' in natural yoghurt results from the lactic acid.

provides the energy source for the starter bacteria, and the milk proteins influence the consistency and texture of the final product, which is known as natural yoghurt. The yoghurt is then cooled to stop the bacteria from growing and turning the yoghurt too sour. Flavouring can be added, such as fruits (which may be fresh fruit pieces or sweetened fruit pulp), honey and nuts. The yoghurt is then packaged and refrigerated.

lactose + bacteria = lactic acid

Yoghurt can be made from a variety of milks, such as from cows, sheep, goats, buffaloes, camels, donkeys, llamas, mares, yaks and even soybeans. The differences in taste and texture depend on the type of milk used. In Australia most yoghurt is made from cow's milk; however more and more people are trying other varieties. Which types of yoghurt have you consumed?

MILK, YOGHURT AND CHEESE

Yoghurt is considered easier to digest than milk because the fat and protein are partially digested during its production and the lactose is broken down into lactic acid.

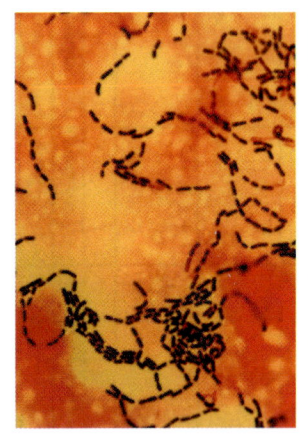

## Properties of Yoghurt

Yoghurt is a semi-solid, creamy curd. Being a dairy product, it contains similar nutrients to that of milk and is therefore an excellent source of calcium, phosphorus, high-quality protein and the B group vitamins.

Yoghurt is often claimed to have health benefits because it contains amounts of healthy bacteria that are usually present in our intestines. Infections or drugs, such as penicillin, may destroy the intestinal healthy bacteria, so eating yoghurt can restore the imbalance.

People who are lactose intolerant are usually able to consume yoghurt as the bacteria have broken down the lactose.

## Classify Yoghurt

**e-fact**

Soft serve or frozen yoghurts are based on low- or reduced-fat ice cream mixes that are flavoured with yoghurt cultures to give it the characteristic yoghurt tang.

The four main types of yoghurt include:
- Full-fat (whole-milk) yoghurt contains not less than 3.2 per cent milk fat and may have added milk powder.
- Reduced-fat yoghurt contains 1–2 per cent milk fat.
- Low-fat (skim milk) yoghurt contains not more than 0.5 per cent milk fat.
- Diet (artificially sweetened) yoghurt contains not more than 0.2 per cent milk fat.
- No-fat yoghurt contains less than 0.15 per cent fat.

## Yoghurt Products

### Vaalia

Vaalia yoghurts are classified as reduced-fat: they contain 1.2 per cent fat. Recently, a range of 'No Fat' yoghurts has been launched. Each 150-gram tub contains a daily dose of acidophilus, bifidus and *Lactobacillus GG*. These bacteria are believed to have a beneficial effect on a person's digestion.

NEED A LITTLE CULTURE?

## ▶ Let's remember

1. Why was yoghurt first created?
2. Why is yoghurt heated during its production?
3. What does the term *coagulation* mean?
4. What does the term *lactose* mean?
5. What causes the slightly acidic taste in natural yoghurt?
6. What do the differences in the taste and texture of yoghurt depend on?
7. List the different types of milk that can be used to make yoghurt.
8. Describe how you would store yoghurt.
9. Identify the nutrients present in yoghurt.
10. Why is yoghurt considered an ideal food for the young or elderly?

## ▶ Let's investigate

1. Visit a supermarket and list all of the different varieties of yoghurt.
2. Survey the class about their favourite yoghurt flavour. Graph your results. To better present your work, you might like to use a computer spreadsheet program.
3. Invent your own yoghurt. What would you name it? What flavour is it? **Design** a label for your yoghurt.
4. Is chocolate yoghurt a healthy snack? Give reasons for your answer. Why do you think it was invented?
5. a  Bring in a variety of yoghurt tubs. Using their nutritional labels, complete the table below for a 100-gram serving of each variety.

| Nutrients | Variety 1 | Variety 2 | Variety 2 |
|---|---|---|---|
| Energy (kj) | | | |
| Protein (g) | | | |
| Saturated fat (g) Total fat (g) | | | |
| Carbohydrates Total (g) Sugars (g) | | | |
| Calcium (mg) As % of RDI | | | |
| Sodium (mg) | | | |

   b  **Evaluate** the nutritional differences between each variety.
   c  Graph the differences in nutrients. To better present your work, you might like to use a computer spreadsheet program.
   d  List the additive numbers found in each variety of yoghurt. Go to the ANZFA website (www.foodstandards.gov.au) and click on the links to the *Food Additives Shoppers' Guide* to find the names of these additives. (They are listed numerically or alphabetically.)

---

**e-DEFINE**

RDI stands for recommended daily intake.

**WEBExtras**

www.vaalia.com.au
The Vaalia website is currently under construction but its aim will be to provide useful information on nutrition and healthy eating.

www.bornhoffen.com.au/about_yoghurt
This Bornhoffen webpage provides information on the health benefits of eating yoghurt.

www.dairycorp.com.au/health/health_yoghurt.htm
The Australian Dairy Corporation's Health and Nutrition website provides comprehensive information on yoghurt, such as its history, its unique properties, the benefits to people's health and how it can help specific medical problems. It also provides some delicious recipes.

MILK, YOGHURT AND CHEESE

6 Yoghurt is used extensively in food preparations, such as sauces, salads and various meat and vegetable dishes. Research a variety of recipes that includes yoghurt as an ingredient. Make a collage of these recipes. Useful resources could be recipe books, websites and magazines.

7 The people from the Balkans consume a refreshing drink called lassi. Find out what is contained in this drink. If you are able to make it in your practical classes, describe its flavour, appearance and texture. Would you consume it again? Why or why not?

## e-Yoghurt

### www.jalna.com.au

1 The word *yoghurt* is of Turkish origin, meaning fermented milk. Find out the French, Bulgarian, Indian and Nepalese terms for yoghurt.
2 Why are bacteria added to yoghurt?
3 Explain why acidophilus, bifidus and *Lactobacillus casei* are considered more beneficial than the bacteria used to start the fermentation process.
4 List the traditional uses of yoghurt in India.
5 Outline what the term *lactose intolerance* means.
6 What are the symptoms of lactose intolerance?
7 Why are lactose-intolerant people more able to consume yoghurt?

### www.pauls.com.au

1 Describe the four main categories of yoghurt. Provide a description for each.
2 Why is yoghurt considered more nutritious than milk?
3 Draw the flow charts to explain the manufacture of natural set and fruit yoghurt.
4 What type of yoghurt is most popular?
5 How is this yoghurt used?
6 Why is plain/natural yoghurt considered to be versatile?
7 How is plain yoghurt used in food preparation?
8 Explain how yoghurt should be stored.
9 What is the shelf-life of yoghurt?
10 How should you thaw frozen yoghurt?

### WEBExtras

**www.dairy.com.au**
This is the Australian dairy industry's website, which serves as a gateway to Australian dairy websites, companies and organisations.

**www.tasdairy.com.au/childrenspages/dairypage.html**
The dairy page of the Tasmanian Dairy Industry Authority can be used for project information.

**www.kraft.com.au**
The Kraft Australia website has a great kids' area, providing lots of games and useful information, such as a virtual museum that details the eating habits of Australians over the past century.

## Puzzled

### Unscrambling milk sugar

Answer the statements to unscramble the letters in the boxes to find another word for the sugar found in milk.

1 Milk and yoghurt are _ ☐ _ _ _ products.
2 This type of yoghurt has no added flavourings: _ _ _ _ _ _ ☐ .
3 Yoghurt is an excellent source of two minerals, calcium and _ _ ☐ _ _ _ _ _ _ .

NEED A LITTLE CULTURE?

4 Yoghurt is also an excellent source of this nutrient: _ _ _ _ ☐ _ _.
5 This is added to cause the yoghurt to thicken: _ _ ☐ _ _ _ _ _.
6 Yoghurt can be made from the milk of this animal: ☐ _ _ _ _.
7 The acid produced when making yoghurt is _ _ _ ☐ _ _.

Answer: _____

### Flavour scramble

Unscramble the following flavourings that are commonly added to yoghurt.

1 sybwaterrr
2 leazhtun
3 yeonh
4 hoocteal
5 tripoca
6 rrrpbsyapes
7 aallvin
8 rryhce
9 sliemu
10 nnaaab

### Nutrient find

Identify some of the nutrients found in yoghurt. Read the following functions and name the nutrient. Good Luck!

1 This nutrient is important in maintaining bone density mass: c _ _ c _ _ m.
2 This nutrient coagulates during the production of yoghurt: _ _ _ t _ _ n.
3 This nutrient works in conjunction with calcium to ossify (or harden) bones: _ h _ s _ _ _ r _ _.
4 This nutrient is a mineral that has a chemical symbol of Mg: _ _ g _ _ s _ _ _.
5 This nutrient is very energy dense: _ _ t.

## WEBExtras

**www.dairypage.com.au**
The Queensland Dairyfarmers' Organisation's DairyPage aims to provide up-to-date information on the Queensland dairy industry.

**www.food.vic.gov.au**
To obtain information on the Victorian dairy industry, first click on 'Victorian Food Industry', then 'Sectors' and finally 'Dairy'.

MILK, YOGHURT AND CHEESE

# Let's Produce

## Make your own yoghurt (serves 1)

### Ingredients
200 millilitres UHT milk
1 tablespoon powdered milk
4 teaspoons natural yoghurt
flavourings

### Method
1. Stir the milk powder into the UHT milk.
2. Warm the milk mixture in a saucepan or microwave until it is approximately 43°C.
3. Add the natural yoghurt and whisk.
4. Pour into plastic cups and cover.
5. Place near a heater for four to six hours.
6. Cool by standing in a bowl of cold water.
7. Add flavourings and refrigerate.

## Lemon and poppy seed muffins (makes 12)

### Ingredients
2 cups flour
1 teaspoon baking powder
1 teaspoon bicarbonate of soda
$\frac{1}{4}$ cup sugar
2 tablespoons honey
2 eggs
$1\frac{1}{4}$ cup natural yoghurt
2 tablespoons milk
65 grams unsalted butter, melted
1 tablespoon grated lemon rind
1 tablespoon poppy seeds

### Syrup
2 tablespoons sugar
2 tablespoons fresh lemon juice
6 teaspoons water

### Method
1. Preheat the oven to 190°C.
2. Grease muffin tins.
3. Sift flour, baking powder and bicarbonate of soda.
4. Combine sugar, honey, eggs, yoghurt, milk, melted butter, lemon rind and poppy seeds in a large bowl.
5. Stir in sifted flour mixture until just combined. Do not overmix.
6. Place the mixture into muffin tins.
7. Bake for fifteen to twenty minutes or until golden brown.
8. Combine lemon juice, water and sugar in a small saucepan and bring the mixture to the boil.
9. Boil for one minute, stirring constantly.
10. Drizzle the syrup over the muffins when you take them out of the oven.
11. Allow the muffins to cool in their tins for about five minutes.
12. Gently turn the muffins out onto a wire rack to cool.

**e-HINT**
The syrup can be made up prior to the lesson and shared by all of the students.

### Questions
1. Why should you not overmix your muffin mixture?
2. Describe three ways you could use to test whether your muffins were cooked?
3. Why should you cool your muffins on a wire rack?

NEED A LITTLE CULTURE?

## Fruity yoghurt pancakes   (serves 2)

**Ingredients**
- ½ cup self-raising flour
- ⅛ teaspoon bicarbonate of soda
- 1 tablespoon caster sugar
- ½ egg, lightly beaten
- ⅓ cup milk
- ⅓ cup vanilla yoghurt
- ½ Granny Smith apple, peeled and grated
- 2 teaspoons butter
- maple syrup
- ice cream

### Method
1. Sift the flour and bicarbonate of soda.
2. Add sugar and make a well in the centre.
3. Combine egg, milk and yoghurt together and mix into dry ingredients. Stir until smooth.
4. Stir in grated apple.
5. Let stand for ten to fifteen minutes for starch granules to soften.
6. Heat butter in a crêpe pan.
7. Pour a quarter cup of batter into the crêpe pan and cook until bubbles appear and burst on the surface.
8. Turn and brown the other side.
9. Remove from the pan and keep warm while you repeat the process with the remainder of the mixture.
10. Serve with maple syrup and ice cream.

**e-HINT**
Dust with one tablespoon of icing sugar and serve with berry sorbet and berries instead of maple syrup and ice cream.

## Fruit and oat cookies   (makes 12)

**Ingredients**
- ¾ cup rolled oats
- ½ cup mixed dried fruit
- ¼ cup brown sugar
- ½ cup wholemeal flour
- ½ teaspoon baking powder
- ½ teaspoon cinnamon
- 1½ tablespoons natural yoghurt
- 1½ tablespoons vegetable oil
- ½ egg, lightly whisked
- ½ ripe banana, mashed

### Method
1. Preheat the oven to 190°C.
2. Line a baking tray with non-stick baking paper.
3. Combine rolled oats, dried fruit and brown sugar in a bowl.
4. Sift flour, baking powder and cinnamon and mix into rolled oats mixture.
5. In a small bowl, whisk the yoghurt, oil and egg until combined.
6. Add the banana to the liquid ingredients and mix thoroughly.
7. Make a well in the centre of the dry ingredients and add the liquid ingredients, using a wooden spoon to combine the mixtures.
8. Place tablespoonfuls of mixture onto the trays, leaving room between each for spreading.
9. Flatten each tablespoonful of mixture with a wet knife.
10. Bake for about twenty-five minutes, or until golden brown.
11. Cool on a tray.

## QUESTIONS

1. Describe how you measured half an egg.
2. Identify the liquid ingredients in the mixture.
3. Identify the dry ingredients in the mixture.
4. Which ingredients would contribute sweetness to the cookies?
5. Why is it important to sift the dry ingredients?
6. **Investigate** other fruit you could use in this recipe.

MILK, YOGHURT AND CHEESE

## Lamb cutlets with yoghurt and couscous

(serves 2)

### Ingredients
- 100 grams natural yoghurt
- 1 tablespoon mango chutney
- ½ teaspoon cumin
- 1 tablespoon fresh coriander
- 4 lamb cutlets
- 125 millilitres chicken stock
- ½ cup couscous
- 4 dried apricots, finely chopped
- 1 tablespoon currants
- 1 tablespoon flaked almonds
- 2 teaspoons olive oil

### Method
1. Combine 50 grams of yoghurt, chutney, cumin and half the coriander in a bowl.
2. Coat the lamb cutlets with the yoghurt mixture.
3. Cover and place in the fridge to marinate for fifteen to twenty minutes.
4. Meanwhile, bring the stock to the boil.
5. Pour the stock over the couscous and stir with a fork.
6. Cover the couscous and set aside for the liquid to be absorbed; this will take about five minutes.
7. Separate the grains of the couscous with a fork and add the apricots, currants, almond and the remainder of the coriander.
8. Heat oil in a frypan and cook the cutlets for two to three minutes on each side, or until cooked.
9. Place the couscous on a plate, arrange the cutlets and top with remaining yoghurt.

## Banana and strawberry shake (serves 2)

### Ingredients
- 250 millilitres milk
- 125 grams strawberries, hulled
- ½ banana
- 100 grams vanilla yoghurt

### Method
Place all of the ingredients in a blender and process until smooth.

## Coconut and yoghurt fruit salad (serves 2)

### Ingredients
- ¼ rockmelon, peeled, seeded and cut into 3-centimetre cubes
- ½ banana, peeled and thickly sliced
- 1 kiwi fruit, peeled and sliced
- 2 slices pineapple, cut into chunks

### Vanilla coconut yoghurt topping
- ¼ cup desiccated coconut, toasted
- 150 millilitres plain yoghurt
- 1 teaspoon vanilla essence
- 1 tablespoon caster sugar
- 2 shakes nutmeg

### Method
1. Combine fruits in a bowl and refrigerate.
2. Preheat the grill and carefully toast the coconut.
3. Combine the toasted coconut, yoghurt, vanilla, sugar and nutmeg.
4. Serve cold fruit salad accompanied by vanilla coconut yoghurt topping.

# Chapter 13

# Say Cheese: Discover Cheese

**e-fact**
Little Miss Muffet sat on her tuffet, eating her curds and whey...

Cheese is thought to have originated in the desert over 4000 years ago. According to legend, an Arabian tradesman was carrying goat's milk while riding on a camel; the milk was in a bag made from a sheep's stomach. When he opened the bag, he discovered that a lumpy substance had formed and was floating in the milky liquid. The lumpy substance is called curd and the milky liquid is whey. This simple process of milk reacting with rennin (found in a sheep's stomach) is how cheese is made.

Today, cheese is made all over the world and is often named after the town from which it comes. In Australia, places such as Bega, King Island and Tilba Tilba are well known for their cheese industry.

MILK, YOGHURT AND CHEESE

## Classify Cheese

Cheese is most commonly classified according to its texture, such as soft cheese (like camembert) or hard cheese (like Parmesan). It can also be classified according to the source of the milk, such as goat's cheese. Feta cheese is an example of a cheese made from goat's milk; however it can also be made from cow's or sheep's milk. It is a soft, white, crumbly Greek-style cheese. Sometimes cheese is classified according to the process of manufacture, such as mould-ripened cheese (like blue-vein cheese).

More recently, we have started to classify cheese according to its fat content. Look at the cheeses in the supermarket and you will notice that the percentage of fat content is often highlighted on labels, such as 50 per cent reduced-fat cheese or 'light cheese'. Make sure *light* means less fat because sometimes it refers to the colour of the cheese.

### Classification of cheese

| Type of cheese | Description | Examples |
| --- | --- | --- |
| Cheddar | Cheddar cheese originated from Cheddar in Somerset, England. It has a firm, close texture. | Cheddar, Colby |
| Hard, grating cheese | Strong-flavoured cheese, usually served grated or shaved. It is most commonly used with pasta dishes. | Parmesan, Romano |
| Fresh or soft, unripened cheese | Moist, soft cheese has a creamy texture and is usually low in fat. | Cottage cheese, ricotta cheese |
| Mould-ripened cheese | Made using a mould. Blue vein has distinctive blue lines and a strong aroma. Camembert has a mould or rind on the outside. | Blue vein, camembert |
| Eye cheese | This smooth cheese has holes (or 'eyes'). | Swiss, havarti |
| Stretch curd cheese | These cheeses are distinctively stringy because the curd has been stretched and shaped in hot water. | Mozzarella, bocconcini |

## Properties of Cheese

**e-fact**
Did you know that approximately 10 litres of milk is needed to make 1 kilogram of cheese?

Nutritionally, cheese is basically just a concentrated form of milk. We often refer to milk as a 'cocktail of nutrients', which means it contains many nutrients and is very good for us. For those of you who are not that fond of drinking milk, cheese is therefore a great alternative. It is an excellent source of protein and calcium, as well as a good source of riboflavin and vitamin A. When cheese is made, the milky liquid (or whey) is usually discarded. Whey contains lactose, so cheese is a good choice for lactose-intolerant people.

## Focus on Riboflavin

**e-define**
Metabolism is the sum of all chemical reactions that keep the body functioning.

Riboflavin is one of the B group vitamins—it is also known as vitamin $B_2$. It is important for healthy skin, nails, hair and eyes. Riboflavin is also important for metabolism and it helps other vitamins, such as vitamin $B_3$ (niacin) and vitamin $B_6$ (pyridoxine), to carry out their functions.

SAY CHEESE

Riboflavin is found in milk, cheese and yoghurt; some breakfast cereals have it added to them. This vitamin is destroyed by sunlight, so milk should not be left sitting out of the refrigerator. Vegemite and Marmite are also very good sources of riboflavin.

All B group vitamins, including riboflavin, are water soluble: they dissolve in water and move through the body quickly. This is why we should consume foods containing these vitamins each day.

# Cheese Products

One of the most popular cheese products in recent years has been cheese slices. When they were first made, cheddar slices were the only variety available. Now you can buy Swiss, edam, havarti and many different fat-reduced varieties.

## WEBExtras

**www.begacheese.com.au**
This is the website for Bega cheese, which is produced in New South Wales.

**www.lactos.com.au**
This is the website for Lactos cheese, which is produced in Tasmania.

**www.redhillcheese.com.au**
The Red Hill cheese site also has a link to the Elizabeth Creek Goat farm, which is an organic producer of goat's milk.

**www.kraftfoods.com.au**
The Kraft website contains information about cheese products and production and osteoporosis.

## QUESTIONS

Look at the Bega Super Slim label and then answer these questions.
1. Who do you think is the target market for this kind of cheese? Explain why.
2. What does the tick on the label indicate?
3. Would you buy this kind of cheese? Why or why not?

## ▶ Let's remember

1. According to legend, who discovered cheese?
2. Describe the meaning of the terms *curd* and *whey*.
3. Name three places in Australia that have cheese named after them.
4. What are some of the ways we can classify cheese?
5. What does the term *light cheese* mean?
6. From where does cheddar cheese originate?
7. Is cheese good food for lactose-intolerant people? Explain.
8. Why is riboflavin important?
9. How much milk is needed to make 1 kilogram of cheese?
10. What does it mean if vitamins are water soluble?

## ▶ Let's investigate

1. Purchase the twelve cheeses referred to in the cheese classification table on page 130, or select twelve other different kinds of cheeses, and have a class cheese tasting. Undertake a sensory analysis and evaluate according to appearance, flavour, texture and aroma. Which cheese did you like the best? Which did you like the least? Rank your preferences from one to twelve. Add the rankings of all students and work out which cheese was voted the most popular in the class and which was the least popular.

MILK, YOGHURT AND CHEESE

2 **Design** a new flavoured cheese spread that you think would appeal to children aged five to nine. What would you call your cheese spread? **Design** your own label to appeal to this age group.

3 **Produce** your own cheese using 125 millilitres of milk, four to five drops of rennet and a pinch of salt. Just heat the milk until it is warm. Test it on your wrist, just as you would if warming a baby's bottle. Add the rennet and stir gently. After twenty to twenty-five minutes, you will notice the milk starting to thicken (curd) and a yellowy liquid (whey) will appear. Place a sieve over the bowl and line with a square of cheesecloth. Pour the mixture into the cheesecloth. The whey will drain through and the curd will remain. Add salt and taste. What does it taste like?

## ▶ e-Cheese

### www.cheese.com

www.cheese.com contains the name of every conceivable kind of cheese in the world. Search 'Cheese by names' or 'Cheese by countries' to discover all about those you have never heard of!

### www.camembert-france.com

Many cheeses are named after the town they originated from, such as camembert, which is named after a town in France. Visit www.camembert-france.com and write a brief report on the history and manufacture of camembert cheese. See whether you can find your own websites on towns that have cheeses named after them.

### www.dairycorp.com.au/cheese

Follow the links to 'All about cheese'. Look up the glossary of cheese terms and write down a list of words that can be used to describe the appearance, flavour, texture and aroma of cheese.

### www.woolworths.com.au/dietinfo/rsa23.asp

1 What are three examples of mould-ripened cheese?
2 How does cheddar cheese vary? Explain.
3 Explain how the fat content varies between different kinds of cheese.

## ▶ Puzzled

### Ch... ch... cheese

Name all of the cheeses you can think of that begin with the letter *C*.

### Cheese match

1 Match each cheese in the first column of the table on the right with its correct country in the second.
2 What kind of cheese would you associate with:
   a pizza
   b cheesecake
   c spaghetti bolognaise
   d Greek salad
   e Welsh rarebit

| Cheese | Country |
|---|---|
| Holland | Neufchatel |
| Switzerland | Mozzarella |
| Norway | Stilton |
| Greece | Danish blue |
| Denmark | Philadelphia cream cheese |
| Italy | Edam |
| France | Gruyère |
| United States | Jarlsberg |

SAY CHEESE

# Let's Produce

## Jacket potato (serves 1)

### Ingredients
1 large baking potato
10 grams butter
1 stalk celery, chopped finely
$\frac{1}{4}$ carrot, grated
1 tablespoon corn
2 tablespoons sour cream
50 grams tasty cheese, grated

### Method
1. Preheat the oven to 220°C.
2. Pierce the potato with a skewer.
3. Place in a microwave on paper towelling for five minutes or until partly cooked.
4. Place the partially cooked potato on oven rungs and bake for thirty minutes or until cooked. Test with a skewer.
5. Cut a cross shape on the potato and squeeze the potato with both thumbs and forefingers.
6. Add butter and then top with celery, carrot and corn.
7. Top with sour cream and cheese.
8. Serve immediately.

### Variations
1. Replace sour cream with low-fat yoghurt.
2. **Design** your own potato with your preferred choice of toppings.

## Berry whip (serves 2)

### Ingredients
1 egg white
2 tablespoons caster sugar
125 grams low-fat ricotta cheese
$\frac{1}{2}$ cup frozen raspberries, chopped
1 tablespoon lemon juice
$\frac{1}{2}$ teaspoon grated lemon rind
extra raspberries, as garnish

### Method
1. Beat egg white until peaks form.
2. Add caster sugar and beat until peaks are firm.
3. Blend ricotta cheese until smooth. Add raspberries, lemon juice and rind.
4. Fold egg white into the ricotta cheese and raspberry mixture.
5. Garnish.
6. Chill in the fridge until ready to serve.

MILK, YOGHURT AND CHEESE

## Spinach and leek quiche (serves 4)

**Ingredients**
- 1 leek, sliced thinly
- 1 cup spinach, blanched
- ½ cup self-raising flour
- 1 cup milk
- 3 eggs
- ⅔ cup cheese, grated

### Method

1. Mix flour, egg and milk together. Stir until just combined.
2. Add spinach, leek and grated cheese.
3. Place into four greased ramekin dishes.
4. Bake at 180°C for twenty-five to thirty minutes.

**e-fact**
Blanching involves partly cooking vegetables by placing them in boiling water.

## Focaccia melt (serves 2)

**Ingredients**
- 1 slab focaccia bread
- ⅓ cup tomato paste
- ¼ cup sun-dried tomatoes, chopped
- ¼ cup pitted black olives, sliced
- ¼ green capsicum, chopped finely
- 6 slices salami
- 100 grams mozzarella cheese, grated

### Method

1. Spread the focaccia bread with tomato paste.
2. Top with sun-dried tomatoes, olives, capsicum and salami.
3. Sprinkle with grated cheese.
4. Bake at 200°C for fifteen to twenty minutes.

## Savoury cheese muffins (makes 6)

**Ingredients**
- 1 cup self-raising flour
- pinch cayenne pepper
- 1 egg, beaten
- 30 grams melted butter
- ¼ cup milk
- ½ cup tasty cheese, grated
- 2 tablespoons chives, chopped

### Method

1. Sift flour and cayenne pepper.
2. Stir in beaten egg, melted butter and milk. Mix until combined.
3. Add cheese and chives.
4. Place into a greased muffin pan.
5. Bake at 200°C for fifteen to twenty minutes.

# Milk, Yoghurt and Cheese: Assessment Task

This assessment task addresses the outcomes HPIP0501, HPIP0502 and TEMA0501 from the Health and Physical Education and Technology Key Learning Areas. The library research can be used as a basis to complete it.

## ▶ Part 1

Research a dietary-related disease caused by a lack of dairy food, such as osteoporosis, rickets or osteomalacia. Include the following information in your report:

1. signs and symptoms of the disease
2. the people who are at risk of developing this disease
3. how this disease can be prevented
4. where people can seek advice on the prevention of this disease
5. **design** an advertisement to assist with the prevention of this disease

The following websites may be useful:
- www.oesteoporosis.org.au
- www.healthybones.com.au
- www.dairy.com.au

## ▶ Part 2

Refer to the website www.moove.com.au to complete these tasks.

1. Find out the number of milk bottles it takes to make a:
    a. standard garbage bin
    b. stormwater pipe
    c. milk crate
2. Outline the steps you should take to recycle milk bottles.
3. Milk cartons are made from renewable resources. Outline what the term *renewable resource* means.
4. What materials are cartons made from?
5. Describe how cartons are recycled.
6. What can one single milk carton be turned into?
7. **Design** a pamphlet, outlining how to recycle plastics and identifying the reasons to recycle plastic.

MILK, YOGHURT AND CHEESE

### Part 3

Read the newspaper article and answer the questions.

## Milking the latest gadget

Your empty milk cartons may one day tell your electronic shopping list to buy a fresh supply.

It may sound like science fiction, but it could soon become reality, according to Berlin engineer Carsten Neiland.

'Imagine the milk carton reporting that it is nearly empty and the next time you go to the supermarket, the shopping cart's integrated chip leads you to the milk refrigerator,' Neiland says.

Neiland, from the Fraunhofer Institute of Reliability and Computer Integration, works in the fields of polymer technology, microelectronics and photonics, whose latest developments have been on display recently at Polytronic 2001 near Berlin.

So-called polymers, experts believe, are tomorrow's favourite material for electronic chips.

Today the electronic industry uses polymers mostly as an isolator and protective coating of the silicon-based chips used in computers, appliances and car eletronics.

However, since specially made polymers can conduct electricity and be moulded precisely for their intended use, they are seen as an ideal future material for electronic components.

'The intelligent household is no longer just a future vision,' Neiland says.

'If you recall the fears that surrounded the millennium changeover, you realise how many household appliances already contain electronic components that could have been affected.'

The fact that milk cartons do not yet incorporate electronics is due to the high cost of components, which are still silicon-based.

But research and development aimed at making low-cost polymer-based chips is now a high priority.

1 According to the article, how may milk cartons tell your electronic shopping list to buy fresh milk?
2 What are polymers mostly used for in today's electronic industry?
3 Why are polymers seen as an ideal future material for electronic components?
4 Why is it that milk cartons today do not incorporate electronics?
5 Why do researchers believe the milk cartons of the future may incorporate electronics?
6 Design a label for a milk carton in the year 2020.

## Section 6

# Meat, Fish, Poultry, Eggs, Nuts and Legumes

**The Australian Guide to Healthy Eating sample serve**

65–100 grams cooked meat or chicken, e.g. ½ cup lean mince, 2 small chops or 2 slices of roast meat
½ cup cooked (dried) beans, lentils, chickpeas, split peas or canned beans
80–120 grams cooked fish fillet
2 small eggs
⅓ cup peanuts or almonds
¼ cup sunflower or sesame seeds

# Chapter 14

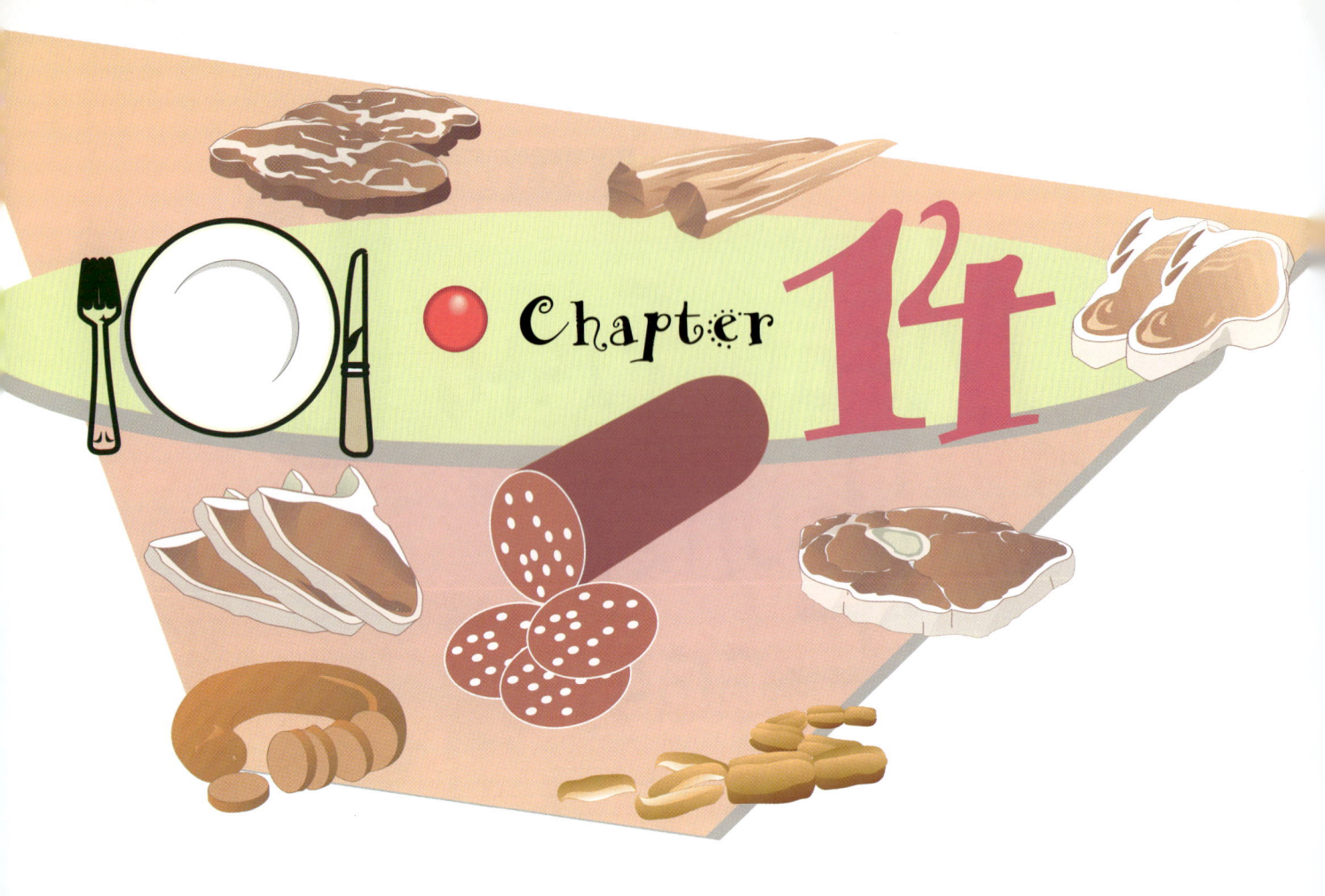

# Steak Your Claim!: Discover Meat

**e-fact**

When the First Fleet arrived in Australia, they brought with them seven cows, a bull, a bull calf, twenty-nine sheep and a few goats, pigs and poultry.

Meat is the edible flesh of domestic animals. According to the Australia New Zealand Food Authority (ANZFA), meat is defined as 'whole or part of

**@-fact**

Did you know that New Zealand has more sheep than people?

the carcass of any buffalo, camel, cattle, deer, goat, hare, pig, poultry, rabbit or sheep, slaughtered other than in the wild state...'

We often talk about meat being red or white: red meat is from cattle, sheep and goats; white meat is from pigs and poultry. Another kind that is becoming more common is game meat, such as venison (deer) and kangaroo.

In Australia and New Zealand, lamb is very popular, more so than in other countries around the world. In North America, beef is very popular, particularly beef hamburgers.

# Classify Meat

We classify meat according to:
- the type of animal it comes from
- the part of the animal that the 'meat cut' comes from

### Types of animals

| Type of meat | Description |
| --- | --- |
| Beef | Meat that comes from cattle |
| Veal | Lean, tender meat that comes from young cattle |
| Lamb | Tender, young meat from sheep aged approximately six months |
| Mutton | Meat from an older sheep |
| Pork | White meat from domestic pigs |

### Types of meat cuts

To make the flesh of animals suitable for consumption, it is prepared into different cuts of meat. These meat cuts come from various parts of the animal and are tough or tender, depending on how much each part is exercised. Tender cuts of meat come from the less exercised parts of the animal, such as the middle (like fillet steak), and tough cuts come from the more exercised parts, such as the shoulder. Tenderness is also related to the age of an animal. Meat from older animals is tougher because the animal has had more exercise.

The type of meat cut also determines the most suitable cooking method. Tough meat should be cooked using long, slow and moist conditions, such as stewing and braising. This enables the meat to become more tender. Tender meat can be cooked more quickly, so grilling and barbecuing are more suitable.

Tender meat is more expensive than the tougher cuts.

MEAT, FISH, POULTRY, EGGS, NUTS AND LEGUMES

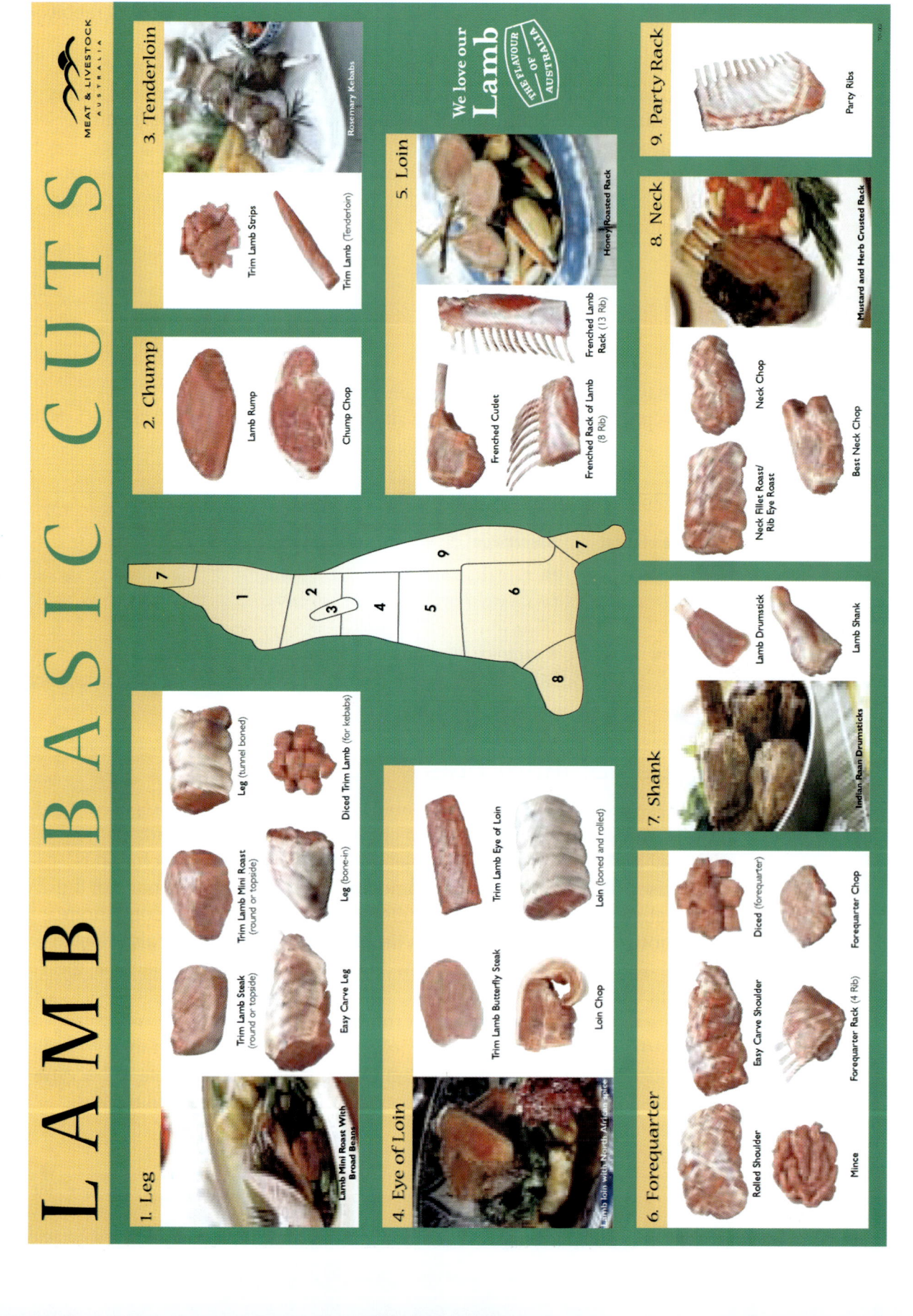

## LAMB BASIC CUTS

MEAT & LIVESTOCK AUSTRALIA

We love our **Lamb** — THE FLAVOUR OF AUSTRALIA

**1. Leg**
- Trim Lamb Steak (round or topside)
- Trim Lamb Mini Roast (round or topside)
- Easy Carve Leg
- Leg (tunnel boned)
- Diced Trim Lamb (for kebabs)
- Leg (bone-in)
- Lamb Mini Roast With Broad Beans

**2. Chump**
- Lamb Rump
- Chump Chop

**3. Tenderloin**
- Trim Lamb Strips
- Trim Lamb (Tenderloin)
- Rosemary Kebabs

**4. Eye of Loin**
- Trim Lamb Eye of Loin
- Trim Lamb Butterfly Steak
- Loin Chop
- Loin (boned and rolled)
- Lamb Loin with a North African spice

**5. Loin**
- Frenched Cutlet
- Frenched Rack of Lamb (8 Rib)
- Frenched Lamb Rack (13 Rib)
- Honey Roasted Rack

**6. Forequarter**
- Mince
- Rolled Shoulder
- Easy Carve Shoulder
- Forequarter Rack (4 Rib)
- Diced (forequarter)
- Forequarter Chop

**7. Shank**
- Lamb Drumstick
- Lamb Shank
- Indian Raan Drumsticks

**8. Neck**
- Neck Fillet Roast/ Rib Eye Roast
- Neck Chop
- Best Neck Chop
- Mustard and Herb Crusted Rack

**9. Party Rack**
- Party Ribs

140

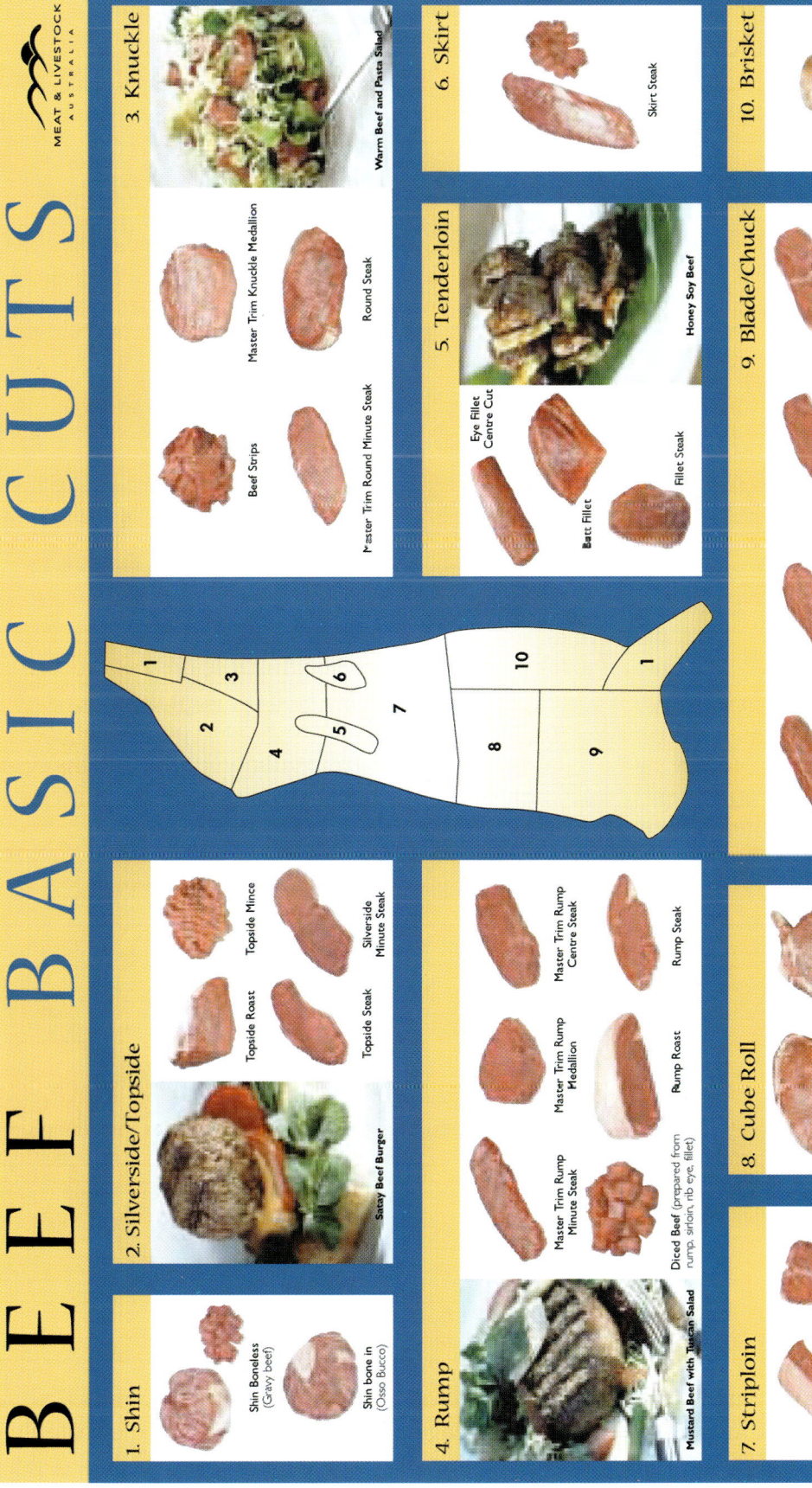

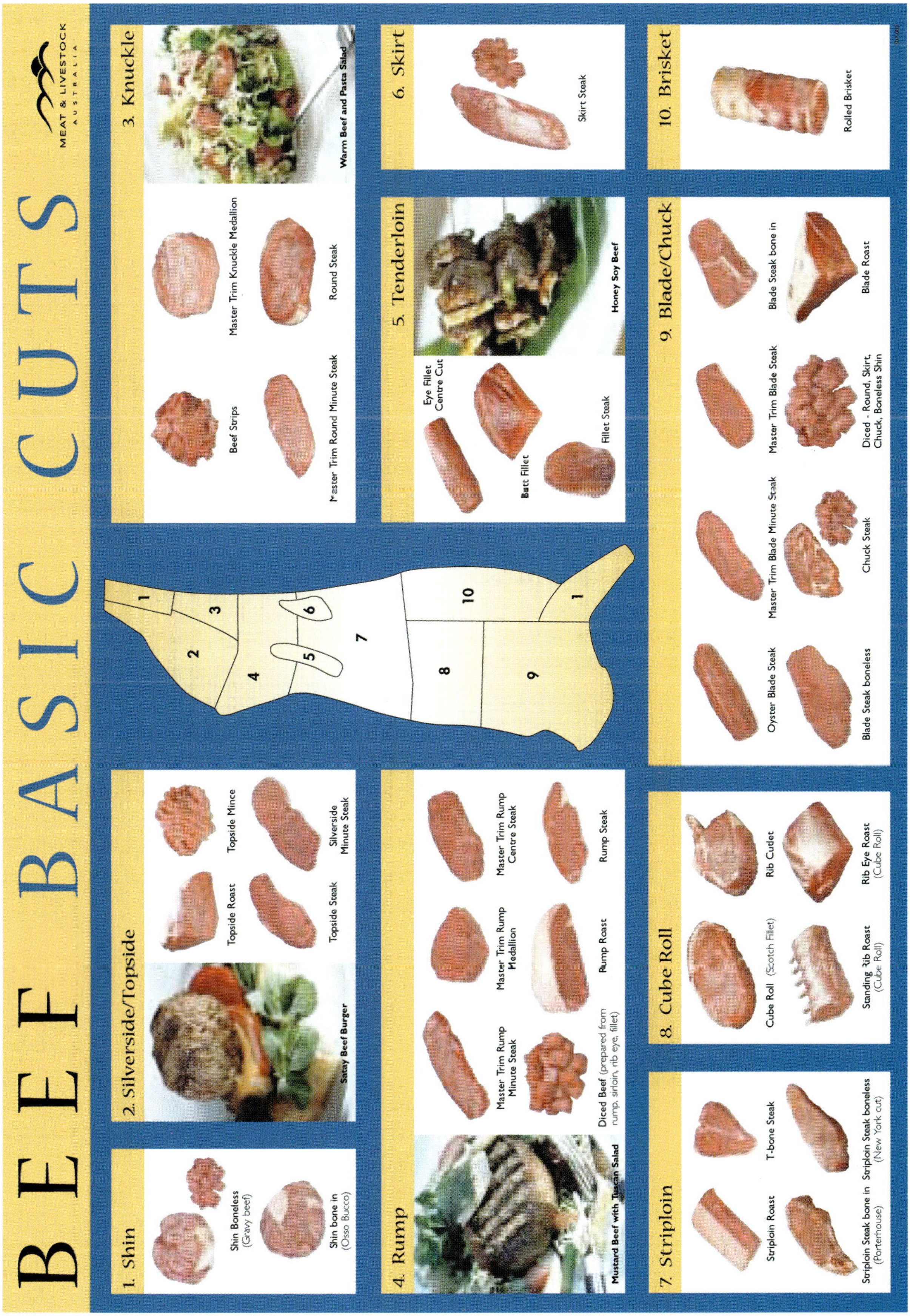

STEAK YOUR CLAIM!

# BEEF BASIC CUTS

MEAT & LIVESTOCK AUSTRALIA

## 1. Shin
- Shin Boneless (Gravy beef)
- Shin bone in (Osso Bucco)

## 2. Silverside/Topside
- Topside Mince
- Topside Roast
- Silverside Minute Steak
- Topside Steak
- Satay Beef Burger

## 3. Knuckle
- Warm Beef and Pasta Salad
- Master Trim Knuckle Medallion
- Round Steak
- Beef Strips
- Master Trim Round Minute Steak

## 4. Rump
- Master Trim Rump Centre Steak
- Rump Steak
- Master Trim Rump Medallion
- Rump Roast
- Master Trim Rump Minute Steak
- Diced Beef (prepared from rump, sirloin, rib eye, fillet)
- Mustard Beef with Tuscan Salad

## 5. Tenderloin
- Honey Soy Beef
- Eye Fillet Centre Cut
- Butt Fillet
- Fillet Steak

## 6. Skirt
- Skirt Steak

## 7. Striploin
- T-bone Steak
- Striploin Steak boneless (New York cut)
- Striploin Roast
- Striploin Steak bone in (Porterhouse)

## 8. Cube Roll
- Rib Cutlet
- Rib Eye Roast (Cube Roll)
- Cube Roll (Scotch Fillet)
- Standing 3 Rib Roast (Cube Roll)

## 9. Blade/Chuck
- Blade Steak bone in
- Blade Roast
- Master Trim Blade Steak
- Diced - Round, Skirt, Chuck, Boneless Shin
- Master Trim Blade Minute Steak
- Chuck Steak
- Oyster Blade Steak
- Blade Steak boneless

## 10. Brisket
- Rolled Brisket

141

MEAT, FISH, POULTRY, EGGS, NUTS AND LEGUMES

# PORK BASIC CUTS

### e-fact

Did you know that new trim lamb steak contains over 80 per cent less fat than traditional chump chops?

As consumers have become more health-conscious over recent years, meat has undergone significant change. We have seen the emergence of such varieties as new trim lamb and new-fashioned pork. Meat also now contains a lot less fat. For example, new-fashioned pork contains about 50 per cent less fat than previously. Because animals are fed differently, their flesh contains less fat.

*Meats of the past (very fatty)*

*Healthier meats of the present (less fat)*

## Properties of Meat

Meat consists of bundles of protein fibres that are held together by connective tissue called collagen. It is a very good source of complete protein, iron and B group vitamins, particularly $B_{12}$.

STEAK YOUR CLAIM!

## ▶ Push for red meat in diet

### Push for red meat in diet

**By TANYA TAYLOR**
**medical reporter**

AUSTRALIANS are eating less red meat because they are confused about its nutritional benefits, an expert committee says.

Its report, *The role of Red Meat in Healthy Diets*, showed Australians ate about half the red meat recommended.

The findings, released yesterday by federal Health Minister Michael Wooldridge, suggested children, women, vegetarians and athletes were missing out on vital nutrients by skipping meat.

Red Meat and Health Expert Advisory Committee chairman Ian Caterson said people should eat lean red meat three or four times a week.

'Lean red meat is a rich source of some very important nutrients like iron, zinc and vitamin B$_{12}$,' Prof. Caterson said. 'It is not a major source of saturated fat or cholesterol.'

Meat and Livestock Australia managing director Richard Brooks denied the timing of the report's release was based on community concern about mad-cow disease. He said it was intended to give Australians correct information about meat.

He said most Australians would be aware the local beef herd was safe from the disease.

Mr Brooks denied the report intended to lean in favor of increased beef and lamb consumption.

Heart Foundation director of health, medical and scientific affairs Prof. Andrew Tonkin said the report showed lean red meat did not increase the risk of obesity or cholesterol.

Dietitians Association of Australian spokeswoman Cathy Cooper warned there were no non-animal sources for vitamin B$_{12}$.

### MEATY FACTS

**WHY IT'S GOOD**
- Protein in beef is 94 per cent digestible, compared with 78 per cent for protein in beans.
- Zinc, an antioxidant, is important for wound healing, bone and hair formation.
- Iron is important for healthy red blood cells and essential for oxygen transport in the body. It is also important for brain development.
- Vitamin B$_{12}$ is found only in animal foods and is important for maintenance of the nervous system.
- Omega-3 fatty acids are needed for health brain and heart function, as well as vision.

**FOR VEGOS**
- Iron is available in green leafy vegetables and legumes, but is not as easily absorbed by the body as the iron in meat.

---

**1** Why is lean red meat recommended?
**2** Why has the community been concerned about red meat consumption?
**3** Explain the roles of protein, iron, zinc, vitamin B$_{12}$ and omega-3 fatty acids.
**4** How much lean red meat should we consume each week?
**5** Does the information in the article come from credible sources? Comment.

## Focus on Iron

Iron is important for the blood as it forms part of the haemoglobin that gives red blood cells their colour. Red blood cells carry oxygen through the blood to other body cells. A lack of iron can cause insufficient amounts of haemoglobin, which means less oxygen can be carried through the blood cells. Anaemia can result from a lack of oxygen in the blood and this can be a problem, particularly for women and adolescent girls. This lack of oxygen causes extreme tiredness and can lead to iron-deficient anaemia, which may require iron supplement tablets. Because women and adolescent girls lose blood each month during their period, they need to ensure they consume foods rich in iron. Vegetarians who do not eat red meat can also be at risk of inadequate iron intake.

Iron can be stored in the liver; however it is important to maintain a steady intake of iron-rich foods. The best sources of iron are liver and kidney. These organ meats are collectively called offal. Red meat is also a good source of iron. Non-meat sources of iron include green leafy vegetables, such as spinach, legumes and cereals.

### e-fact

**Tripe is the inside lining of the stomachs of sheep, pigs and cattle. Did you know there is a tripe club in Sydney, where people indulge in cooking tripe dishes?**

MEAT, FISH, POULTRY, EGGS, NUTS AND LEGUMES

# Meat Products

**e-DEFINE**

To cure something is to preserve it through various methods, such as salting, drying or smoking.

Sausages are the most common type of meat product. Traditionally, they used to be made from pork mince, with added breadcrumbs and preservatives; nowadays, many different varieties are available, such as Thai chicken, Mexican or satay.

Meat is also used to make a variety of meat products, such as ham and bacon. We refer to these products as smallgoods. Ham is a leg of pork that has been cooked and cured, whereas bacon is a side of pork that has undergone curing and wood-fired smoking. Other smallgoods include frankfurts, salami and strasbourg.

## ▶ Let's remember

1. Give an example of red, white and game meat.
2. In which countries is lamb popular?
3. How do we classify meat?
4. What is the difference between mutton and lamb?
5. Explain why some meat is tender and some is tough.
6. Why are different methods of cooking suitable for different cuts of meat?
7. How is today's meat different from what was available in the past?
8. Why is iron so important?
9. Which foods are high in iron?
10. Name three meat products.

## ▶ Let's investigate

1. Find out the meaning of these meat-related words:
   a collagen    b gelatin    c medallion
2. Visit www.colesonline.com.au and find six different kinds of meat, ranging from the cheapest to the most expensive. Which part of the animal do the most expensive cuts come from?
3. Identify as many edible organ meats as you can.

## ▶ e-Meat

www.australianlamb.com.au

The Meat and Livestock Australia website has information on Australian lamb and beef. Visit www.australianlamb.com.au/content.cfm?sid=76 and then answer these questions.

1. Who was the pioneer responsible for bringing merino sheep to Australia?
2. Dairy and beef cattle originally grazed together. What kind of land became more suitable for each?
3. What event in 1879 revolutionised the meat export industry?

STEAK YOUR CLAIM!

#### www.pork.gov.au

1. What is meant by the statement 'Pork is nutrient dense?' (www.pork.gov.au/recipes6.htm)
2. What is the nutritional value of pork? (www.pork.gov.au/recipes6.htm)
3. How should fresh pork be cooked? (www.pork.gov.au/Recipes/recipe72.htm)

#### www.kangaroo-industry.asn.au

Visit the Kangaroo Industries Association of Australia website and then complete these tasks.

1. Follow the link to recipes. Kangaroo meat is very low in fat. Given this, what suggestions are made when cooking kangaroo meat?
2. Follow the link to products. How does the nutritional value of kangaroo meat compare with other types of meat. Graph this information.

## ▶ Puzzled

### Meat search

Find the meat-related words in each sentence. For example:
*Please drop the parcel* **off al***ong the way.*

1. I hope you have a lovely time.
2. The haven is on the island.
3. The cabinet is teak on the outside.
4. *Kanitchiwa* means hello in Japanese.
5. Michael ambled along the beach.

### Meat match

Match the words in each column to make new meat-related words.

| | |
|---|---|
| Hind | Shank |
| Silver | Steak |
| Sir | Side |
| Rump | Quarter |
| Lamb | Loin |

### Country match

Which country do you associate with these dishes:

1. roast beef
2. spaghetti bolognaise
3. samosas
4. Irish stew
5. moussaka
6. chilli con carne
7. beef stroganoff
8. Wiener schnitzel
9. hamburger
10. haggis

### WEBEXTRAS

www.meat4health.com.au
The Meat and Livestock Australia website provides health information on meat.

www.ddbacon.com.au
The KR Darling Downs website provides information on their ham, bacon and other smallgoods products.

www.georgewestonfoods.com.au/brands/donsmallgoods.htm
The George Weston Foods website has a range of information about Don Smallgoods products.

MEAT, FISH, POULTRY, EGGS, NUTS AND LEGUMES

# Let's Produce

## American cheeseburger (serves 1)

### Ingredients
- 150 grams beef mince
- 2 teaspoons vegetable oil
- ½ small onion, sliced thinly into rings
- 1 cheese slice
- 1 hamburger bun
- 1 gherkin, sliced thinly
- 2 teaspoons American mustard
- 2 teaspoons tomato ketchup

### Method
1. Shape the meat into a round shape. Flatten between two small pieces of baking paper.
2. Lightly fry onion in oil until brown. Place to one side of the frypan.
3. Add meat burger and cook each side for approximately five minutes or until cooked through.
4. Place cheese slice on top of meat to melt slightly.
5. Toast hamburger bun lightly on inside.
6. Place burger onto one side of bun.
7. Top with onion, gherkin, mustard and tomato ketchup.
8. Place other half of bun on top. Eat!

### Variations
Design and produce an Aussie hamburger by replacing the gherkin and mustard with lettuce, beetroot, sliced tomato, tomato sauce and maybe even a fried egg and a slice of pineapple!

## Pork and prunes (serves 2)

### Ingredients
- 200 grams pork fillet, sliced thinly
- 1 tablespoon flour
- 1 teaspoon red currant jelly
- ½ cup chicken stock
- 1 teaspoon lemon juice
- ¼ cup cream
- 6 pitted prunes, chopped
- 1½ cups rice, cooked
- 1 tablespoon parsley

### Method
1. Place the flour and pork into a plastic bag. Shake until the flour coats the pork.
2. Mix red currant jelly, stock, lemon juice and cream together.
3. Place liquid mixture, pork and prunes into microwave-safe dish.
4. Cook in microwave on high for seven to ten minutes, or until pork is cooked.
5. Serve with rice and garnish with chopped parsley.

## Veal schnitzel (serves 1)

### Ingredients
- 100 grams veal (whole)
- 1 tablespoon flour
- ½ egg, beaten
- 1 tablespoon milk
- ¼ cup breadcrumbs
- 2 tablespoons oil

### Method
1. Flatten veal with a meat mallet.
2. Coat the veal with flour. Shake off excess.
3. Mix milk and egg together.
4. Dip it into egg and milk mixture.
5. Coat with breadcrumbs.
6. Heat oil in a frypan. Add veal and cook each side for five minutes.

**e-HINT**

Place baking paper into the frypan before adding the oil. Baking paper prevents the veal from losing its coating and sticking to the pan.

STEAK YOUR CLAIM!

## Beef satay (serves 2)

**Ingredients**
200 grams rump steak, chopped
1 teaspoon soy sauce
1 teaspoon honey
¼ teaspoon chilli powder
¼ teaspoon curry powder
1 tablespoon oil
rice
satay sauce

### Method
1. Mix together all of the ingredients and marinate for as long as possible.
2. Soak four skewers in water for ten minutes.
3. Thread meat onto skewers and place under a preheated griller.
4. Cook, turning once for approximately ten minutes, or until meat is cooked through.
5. Serve with rice and satay sauce.

### Variations
Use chicken or pork instead of beef.

## QUESTIONS

1. **Investigate** which countries satays are popular in.
2. Why do the skewers need to be soaked in water?
3. Why should the griller be preheated?

## Chilli con carne (serves 2)

**Ingredients**
100 grams beef mince
2 teaspoons oil
½ onion, chopped finely
¼ green capsicum, chopped finely
1 teaspoon paprika
¼ teaspoon chilli powder
½ cup tomato soup
¼ cup kidney beans
½ cup brown rice

### Method
1. Place rice into a large amount (at least 1 litre) of boiling water and cook for thirty minutes.
2. Heat oil in a pan, add mince and cook gently until browned.
3. Add onion, paprika and chilli powder. Cook for one to two minutes.
4. Add capsicum and tomato soup.
5. Cook for twenty to thirty minutes.
6. Stir in kidney beans.
7. Serve immediately with rice.

# Chapter 15

# Sounds Fishy: Discover Seafood

Did you know that there are over 24 000 species of fish in the world and over 4000 in Australia? As an island continent, Australia is well known for its range of fish and shellfish found in its many rivers and oceans. Over 400 species are commercially harvested and marketed.

What do you think of when you think of fish? Ask some adults about their recollections and they might tell you about how fish and chips used to be sold wrapped up in old newspapers. With changes to food safety laws, fish cannot be sold like this any more. Start thinking about how you can buy fish now, as well as all of the places you can buy fish and all of the ways it is sold. We can buy seafood from the market, the supermarket, speciality seafood shops, fast food outlets, fish and chip shops and restaurants. We can buy fish whole, in fillets or as cutlets, frozen or fresh, in packets, crumbed or even in a sauce.

Seafood can be cooked in many ways and takes little time to prepare. It can be cooked using the following methods: grilling, poaching, steaming,

SOUNDS FISHY

**e-DEFINE**

A moist cooking method is when the food is cooked in liquid.

baking, shallow-frying, deep-frying, smoking or microwaving. It is important to remember that seafood should not be overcooked as the flesh is tender. To maintain the natural juices when cooking, it is advisable to either use a moist method of cooking or baste frequently if using a dry method.

## Classify Fish

Fish can be classified as white, oily or shellfish. White fish contains low levels of fat, examples of which include John Dory, whiting, perch and ling. Oily fish contains higher levels of fat, such as salmon, sardines, mackerel and trout. There are two types of shellfish: molluscs and crustaceans. Molluscs, such as scallops, clams and mussels, have a shell that opens when cooked. Crustaceans, such as prawns, lobster and crab, also have a hard outer covering or shell that turns pink, orange or red when cooked.

## Properties of Fish

It is recommended that we should eat seafood at least once a week. A 150-gram serving provides approximately 50–60 per cent of the recommended daily intake (RDI) for protein for an adult. Not only is fish high in protein, it is also low in fat.

Seafood is also an excellent source of vitamins and minerals, especially iodine, which enables the thyroid gland to function properly. It also contains vitamin A, vitamin D, niacin, iron and zinc. Some shellfish, such as oysters, mussels and scallops, are a very good source of iron and zinc; canned fish, like salmon, contains soft, edible bones—an excellent source of calcium.

MEAT, FISH, POULTRY, EGGS, NUTS AND LEGUMES

## Focus on Omega-3

**e-DEFINE**
Polyunsaturated fats have two or more double bonds in the linked carbon chain. Each double bond is able to admit two hydrogen atoms. They are not associated with the accumulation of cholesterol.

Much research has been done recently on omega-3 fats. While eating low-fat food is considered to be an important part of healthy eating, we now know that some fats are especially good for us. These are called omega-3 fats; they are found in most types of seafood. Such polyunsaturated oils are thought to reduce the risk of heart disease, stroke, asthma, arthritis, multiple sclerosis, psoriasis, inflammatory bowel disease, cancer and even Alzheimer's disease.

If you are allergic to seafood or dislike eating it, omega-3 fats can be obtained from other sources, such as:
- canola oils and margarine spreads
- soybeans
- linseed, which is found in many cereal products, like breakfast cereals and breads
- omega-3 eggs
- green vegetables, such as spinach, peas and beans

**e-DEFINE**
Psoriasis is a severe skin condition with red patches and white, flaky lesions.

## Seafood Products

Consumers purchase products for a variety of reasons. For example, canned tuna has been a product available on supermarket shelves for many years. It is tasty, convenient to use, economical, versatile and a good source of nutrients. Some consumers also consider environmental issues when buying particular products. Tuna manufacturer Greenseas displays the 'Dolphin Safe' symbol, indicating that the tuna has not been caught in dolphin zones or by such methods as drift-nets. This guarantee that innocent dolphins are not killed is very important to some consumers.

### ▶ Let's remember

1. Why do we have so many varieties of seafood in Australia?
2. Name three ways fish can be sold.
3. Why is it important not to overcook fish?
4. How is fish classified?
5. How do molluscs and crustaceans differ?
6. What types of fish contain calcium?
7. What are omega-3 fats?
8. Besides seafood, what other types of food are good sources of omega-3 fats?
9. Name one dry and one moist method of cooking seafood.
10. What can influence consumers when choosing products?

SOUNDS FISHY

## ▶ Let's investigate

| Food source | mg per 100 g |
|---|---|
| Fish | 210 |
| Oysters | 150 |
| Prawns | 120 |
| Lobster | 105 |
| Turkey | 35 |
| Beef | 22 |
| Chicken | 19 |
| Lamb | 18 |
| Pork | 0 |
| Veal | 0 |

Source: CSIRO research (found at www.siv.com.au/nutrition.html)

1 Using the information in the table on the left, graph the amounts of omega fatty acids present in different foods. To better present your work, you might like to use a computer spreadsheet program.

2 **Investigate** what to look for when buying fresh fish. Then complete the table below. The first row has been done for you.

| Eyes | Eyes should be clear, not cloudy and bulging, not sunken. |
|---|---|
| Flesh | |
| Skin and scales | |
| Odour | |
| Colour | |

3 Choose a fish and write a short story about its life in the ocean or river.

4 **Investigate** the terms *caviar* and *roe*.

## ▶ e-Fish

### www.chsmith.com.au/fish-prices/MWFMarket

Visit the Melbourne Wholesale Fish Market and make a list of the different types of freshwater fish, saltwater fish and shellfish. Select one that is suitable for these four recipes at the end of this chapter: sweet curried fish, crispy baked fish, Indian fish curry and spaghetti marinara. Look up the prices of your chosen fish and work out how much the fish would cost for each of the recipes.

### www.siv.com.au/cooking.html

Many types of seafood are available all year round, whereas some are seasonal. Visit the Seafood Industry Victoria website and find out the seasonal availability for flounder, snapper, rock lobster and scallops. What methods of cooking are suggested for these types of fish? Some species are commonly mislabelled. What are the correct names for crayfish, sea perch and trevally?

### www.greenseas.com.au

1 Visit the Heinz Greenseas website and **investigate** the range of tuna products available. Choose three products and research their:
   a physical properties—describe the taste, texture and appearance
   b chemical properties—what nutrients does it contain?
2 Why might each product appeal to consumers?
3 Search the 'Recipe Ideas' section to find a recipe using each product.
4 Using one of the recipes you have chosen, **design** your own by altering at least 10 per cent of the original. Think about ingredients you could substitute and how quantities could be varied.

MEAT, FISH, POULTRY, EGGS, NUTS AND LEGUMES

## Puzzled

### Drop in the ocean

Find the twenty-nine types of seafood from the box in the grid below. Can you then find the hidden message?

| Balmain bug | barramundi | blue grenadier | bream | calamari | clam |
| cod | crab | crayfish | flake | flathead | flounder |
| gemfish | hake | John Dory | ling | mackerel | mussel |
| oyster | perch | pipi | prawn | salmon | sardine |
| scallop | snapper | trevally | tuna | whiting | |

| G | E | M | F | I | S | H | R | C | S | C | E | A | P | I |
|---|---|---|---|---|---|---|---|---|---|---|---|---|---|---|
| M | F | F | O | O | D | I | E | S | R | R | A | N | I | R |
| B | U | M | L | G | E | N | I | D | R | A | S | F | P | A |
| E | R | S | A | A | U | X | D | C | C | Y | B | L | I | M |
| Y | E | E | S | C | T | B | A | L | L | F | L | A | E | A |
| L | N | T | A | E | K | H | N | S | A | I | O | K | U | L |
| L | R | C | R | M | L | E | E | I | M | S | H | E | E | A |
| A | H | C | R | E | P | O | R | A | A | H | F | A | P | C |
| V | L | I | N | G | D | R | G | E | D | M | O | T | K | R |
| E | S | A | L | M | O | N | E | E | L | I | L | N | O | E |
| R | E | T | S | Y | O | L | U | A | N | U | T | A | K | P |
| T | P | R | A | W | N | D | L | O | T | S | D | N | B | P |
| W | H | I | T | I | N | G | B | G | L | R | O | D | T | A |
| B | A | R | R | A | M | U | N | D | I | F | C | D | D | N |
| Y | R | O | D | N | H | O | J | P | O | L | L | A | C | S |

### Hook, line and sinker

Answer true or false to these statements.
1  Brine refers to a solution of salt and water.
2  Another name for sardines is pilchards.
3  Flake is a type of shark.
4  Yabbies are found in the ocean.
5  Lobsters turn purple when they are cooked.
6  Whitebait is a small fish that is eaten whole.
7  Seafood is a very good source of iodine.
8  In Japanese society, fish is often eaten raw.
9  Another name for calamari is squid.
10 Prawns are an example of a mollusc.

SOUNDS FISHY

# Let's Produce

## Sweet curried fish  (serves 2)

### Ingredients
300 grams potatoes
2 tablespoons curry powder
2 tablespoons sugar
2 fillets fish (approximately 150 grams each)
3 teaspoons butter
2 tablespoons milk
1 tablespoon parsley, chopped
sprigs of parsley or coriander, as garnish

### Method
1. Peel and cut potato into even-sized pieces. Cook in boiling water for twenty minutes or until tender.
2. Combine curry powder and sugar and then coat both sides of fish with mixture.
3. Melt two teaspoons of butter in a frypan and gently cook each side of the fish for five to seven minutes, turning carefully.
4. Drain potatoes and mash, adding one teaspoon of butter and two tablespoons of milk. Mix in parsley.
5. Serve the fish on a bed of potato mash. Garnish with a sprig of parsley or coriander.

**e-HINT**
Coat fish just before cooking as soggy fish makes the coating wet.

## Tuna fish cakes  (serves 4)

### Ingredients
175 grams pontiac potatoes
150 grams flaked tuna (Greenseas Lemon and Cracked Pepper Tuna in Spring Water)
½ egg, lightly beaten
½ cup breadcrumbs (use 1-day-old bread)
1 shallot, finely chopped
1 teaspoon lemon juice
1 tablespoon parsley, chopped
1 teaspoon Dijon mustard
1 tablespoon oil
75 grams salad greens
½ lemon, cut into wedges

### Method
1. Cook potatoes in boiling water for twenty minutes or until tender. Drain and mash.
2. Add tuna, egg, breadcrumbs, shallots, lemon juice, parsley and mustard to mashed potato, mixing until well combined.
3. Divide mixture into four even portions. Using wet hands, shape into an oval shape. Pinch one end of the oval to make the shape of a fishtail.
4. Place fish cakes on baking tray lined with lightly greased baking paper. Place in fridge and chill for ten to fifteen minutes or until firm.
5. Heat oil in a non-stick pan. Add fish cakes and cook each side for five minutes, or until heated through and golden brown on outside.
6. Serve on a bed of salad greens and garnish with lemon wedges.

MEAT, FISH, POULTRY, EGGS, NUTS AND LEGUMES

### Ingredients

1 cup non-crinkle potato crisps, finely crushed
2 tablespoons Parmesan cheese
1 tablespoon parsley, chopped
2 tablespoons lemon juice
¼ cup mayonnaise
2 fillets fish (approximately 150 grams each)
2 tablespoons flour

## Crispy baked fish (serves 2)

### Method

1. Preheat oven to 180°C.
2. Combine crisps, Parmesan cheese and parsley.
3. Combine lemon juice and mayonnaise.
4. Lightly coat the fish with flour and then dip in mayonnaise mixture. Coat fish with potato chip mixture.
5. Place the fish in a baking dish lined with lightly greased baking paper. Bake in the oven for ten to fifteen minutes, or until fish is cooked.

### Ingredients

15 grams butter
½ brown onion, chopped finely
1 clove garlic, crushed
1 teaspoon ground cumin
1 teaspoon ground coriander
1 teaspoon garam masala
¼ teaspoon turmeric
½ teaspoon chilli powder
200 grams canned tomatoes, chopped
100 millilitres coconut milk
200 grams fish fillets
½ cup basmati rice
coriander, for garnish

## Indian fish curry (serves 2)

### Method

1. Melt butter in a large saucepan. Add onion and cook for two to three minutes. Add garlic, cumin, ground coriander, garam masala, turmeric and chilli powder. Cook for one minute.
2. Stir in tomatoes and coconut milk. Reduce heat and simmer for five minutes. Add fish and simmer for a further five minutes or until fish is just cooked.
3. Serve fish on a bed of rice, garnish with coriander sprigs.

### QUESTIONS

1. Garam masala is a blend of ingredients. **Investigate** what these are by either looking at the label on the container or referring to a reference book about food and ingredients.
2. **Investigate** the origin of turmeric and the kinds of dishes you would you use it in. (Use labels and reference books to assist you.)
3. **Investigate** the difference between coconut milk and coconut cream.

SOUNDS FISHY

## Spaghetti marinara (serves 2)

**Ingredients**
200 grams spaghetti
1 tablespoon olive oil
¼ brown onion
1 clove garlic
200 grams chopped tomatoes, canned
60 millilitres vegetable stock
1 tablespoon tomato paste
150 grams firm fish fillets, diced into 2-centimetre cubes
4 prawns (remove veins and shell, leaving tail intact)
4 scallops
1 tablespoon cream (optional)
2 teaspoons chopped basil

**Method**

1. Cook spaghetti in a large saucepan of boiling water until al dente.
2. Heat oil in a frypan. Add finely chopped onion and garlic. Cook for two to three minutes.
3. Add tomatoes, stock and tomato paste. Bring to the boil and then simmer for five minutes.
4. Add fish and prawns to sauce. Cook gently for five minutes.
5. Add scallops and simmer gently until all seafood is just cooked.
6. Drain spaghetti and return to the saucepan. Add seafood sauce mixture. Combine all ingredients, tossing gently. Add one tablespoon of cream, if desired.
7. Serve immediately, sprinkling with basil.

## QUESTIONS

1. What does *al dente* mean?
2. Some people add a small amount of oil to the water when cooking pasta. Why do you think this would be done?
3. **Investigate** the difference between the terms *boil* and *simmer*.

# Chapter 16

# One for the Birds: Discover Poultry

The term *poultry* refers to the meat obtained from a range of farmyard-bred birds, such as chickens, geese, ducks and turkeys.

### e-fact
Did you know that Donald Duck lives at 1313 Webfoot Walk, Duckburg, Calisota?

ONE FOR THE BIRDS

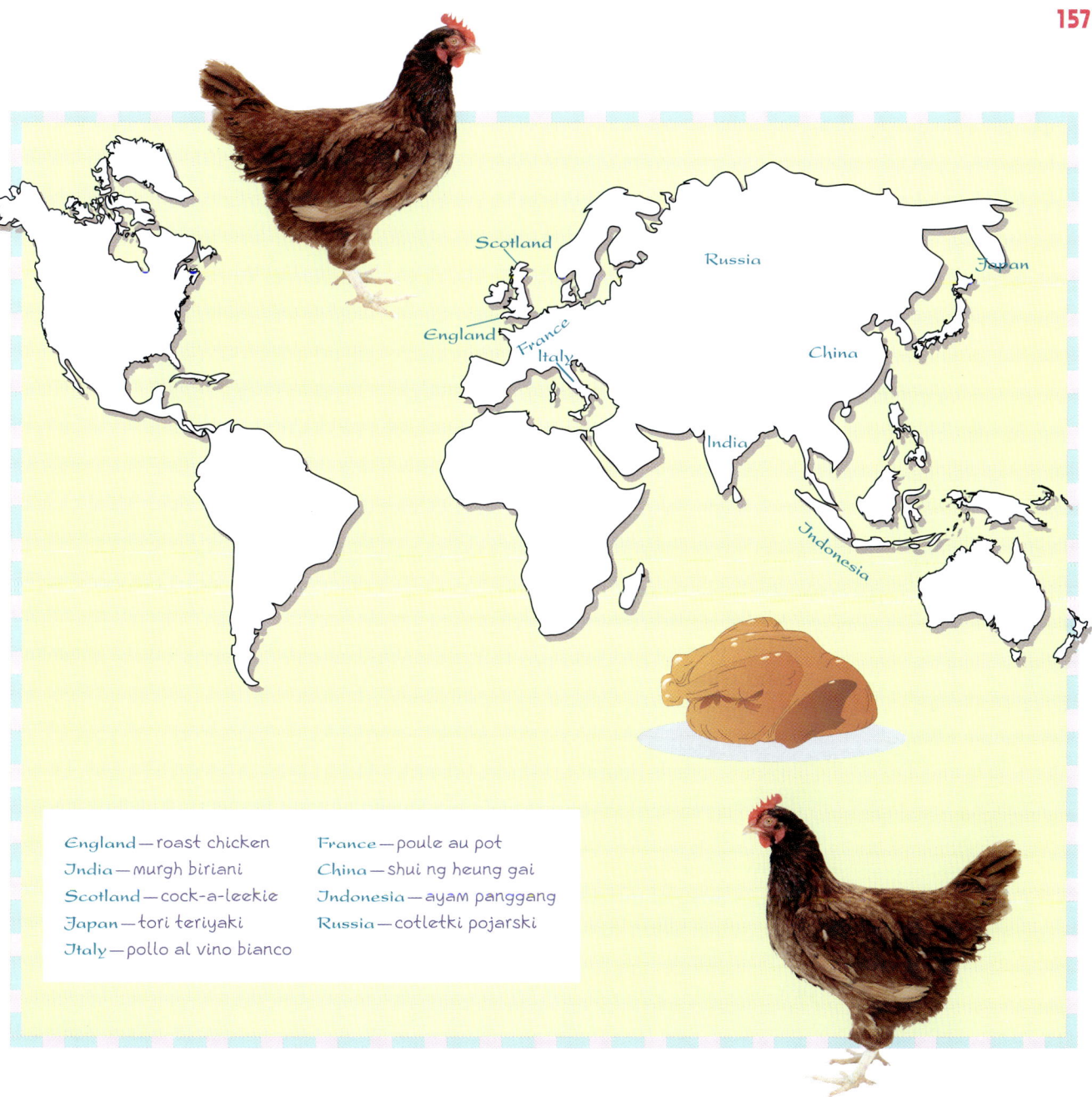

157

England — roast chicken
India — murgh biriani
Scotland — cock-a-leekie
Japan — tori teriyaki
Italy — pollo al vino bianco
France — poule au pot
China — shui ng heung gai
Indonesia — ayam panggang
Russia — cotletki pojarski

### e-fact

**In 1950 three million chickens were processed compared with almost three times this amount today.**

In Australia, chicken is the most common type of poultry consumed; in fact, it is one of the world's most popular meats. Chicken descended from the wild red jungle fowl of India, and is now part of many traditional dishes of various countries.

Prior to the 1970s chicken was considered to be a luxury food because it was only consumed on special occasions; it was often the sought-after first prize in raffles at the local pub. However the use of selective breeding practices and antibiotics has meant that it is much cheaper and quicker to produce today. Contrary to public opinion, growth hormones are not used in the production of poultry in Australia; in fact, it is has been banned since the 1960s.

MEAT, FISH, POULTRY, EGGS, NUTS AND LEGUMES

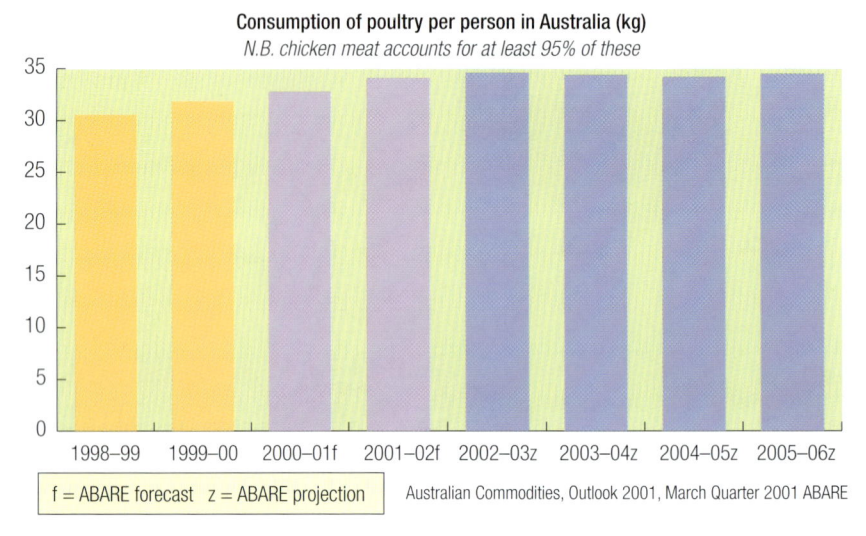

**e-RIDDLE**

Q: What do chickens grow on?
A: Eggplants

**Consumption of poultry per person in Australia (kg)**
*N.B. chicken meat accounts for at least 95% of these*

f = ABARE forecast   z = ABARE projection    Australian Commodities, Outlook 2001, March Quarter 2001 ABARE

Initially, chicken was purchased in the frozen form; however today, owing to consumer demand, it is purchased in a variety of forms:
- frozen chicken pieces
- chilled fresh whole chicken
- chilled fresh chicken portions, such as mince, fillets, thighs, drumsticks and wings
- pre-prepared chicken dishes, such as those from Lenard's (a franchise of stores that retails fresh, pre-prepared poultry, lamb, pork and beef products), delicatessens and supermarkets
- frozen chicken products, such as nuggets and schnitzels

Not only has the expansion of takeaway food outlets, such as KFC and Red Rooster, contributed significantly to the increase in chicken sales but also because of a belief that white meat was healthier than red meat. Though the latter has mainly been caused by a misunderstanding about the nutritional differences between red and white meat—if selected, prepared and cooked correctly, both can be nutritious.

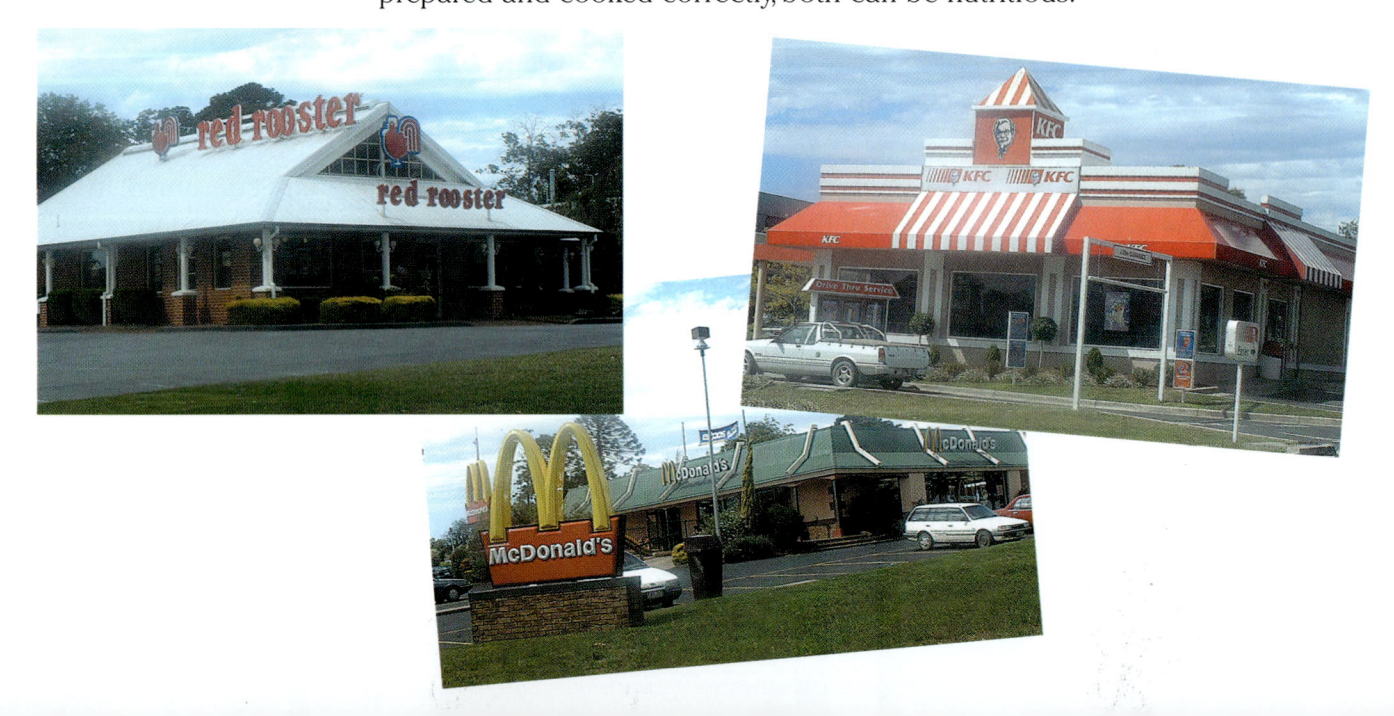

ONE FOR THE BIRDS

## e-RIDDLE

Q: How do chickens dance?
A: Chick to chick

The chicken industry responded well to consumer demand for convenient, quick-to-prepare meals with a range of cuts. These have been effectively promoted along with recipe suggestions to suit our busy lifestyles.

There is much controversy about chicken production. Chickens are often mass-produced on poultry farms, where they are caged, fed a specially formulated diet and slaughtered once they have reached the desired weight. Many people believe this farming method is cruel because the chickens have limited mobility and their quality of life appears quite minimal. However free-range methods are gaining in popularity, where the chickens are raised in smaller flocks and allowed to roam free on farms. The resulting chicken meat tends to be more expensive.

Fresh chicken can be stored for two to three days completely covered in the refrigerator. Frozen chicken should be thawed completely before being cooked. Raw chicken is subject to salmonella, a dangerous bacterium that causes food poisoning. This means chicken must be thoroughly cooked before being eaten; cooking also tenderises the flesh and develops flavour.

## Properties of Chicken

Chicken is a white meat and so offers a subtle, mild flavour. It is an excellent source of complete protein and a reasonably good source of iron—it does not provide the same amount of iron as red meat. If the skin is removed, chicken is low in fat; most fat is found in or just under the skin.

MEAT, FISH, POULTRY, EGGS, NUTS AND LEGUMES

## Classify Chicken

Chicken is classified according to weight. Australia uses a numerical system, where each size increases by 100 grams:
- A chicken labelled size 5 will weigh 500 grams.
- A chicken labelled size 6 will weigh 600 grams.
- A chicken labelled size 11 will weigh 1.1 kilograms.
- A chicken labelled size 12 will weigh 1.2 kilograms.

What will a chicken labelled size 14 weigh?

## Chicken Products

### Case Study: Lenard's

Lenard's is a network of over 200 stores located in Australia, New Zealand, South Africa and Ireland. Founded in 1987, Lenard's Poultry Shop is a unique concept of value adding to fresh produce. Its name was changed to Lenard's in 1998 when other meats, such as lamb, pork and beef, were added.

The founder of Lenard's was Lenard Poulter, a Melburnian butcher. Poulter could foresee the changes in working patterns—consumers would be working longer hours and leading more hectic lifestyles. Poulter also realised that these changing social trends would not mean that consumers would want to compromise the quality or freshness of their products, nor would they want to pay a lot more for value-added products. With the growing trends for poultry, he opened his first store in 1987.

So much has happened since then:
- Lenard's has served more than 8.7 million customers in Australia.
- Every week Lenard's has about 168 000 customers in Australia.
- Lenard's turnover increased from $5 to $85 million from 1989 to 2000.
- In 1998 Lenard's used 800 000 fresh chickens in its products; in 2000, over 6.5 million.

Poulter believes the key to the success of Lenard's belongs to several key elements:
- freshness
- convenience
- value
- service
- comprehensive cooking guides, brochures and great knowledge of the product
- quality

All of Lenard's products are prepared in store every day.

### Questions

1. Outline what types of meats Lenard's sells.
2. Why was Lenard's developed?
3. What is the key to Lenard's success?
4. Describe some of the products sold at Lenard's.
5. **Design** a new poultry dish that could be sold at Lenard's. Describe the dish's ingredients in detail and draw a diagram of your creation. Ensure you give it a creative name.
6. Visit Lenard's website at www.lenards.com.au to find out more about this franchise.

ONE FOR THE BIRDS

## ▶ Let's remember

1 Identify five animals that are classified as poultry.
2 Identify the traditional chicken dish for these countries:
   a England        d Japan         g China
   b India          e Italy         h Indonesia
   c Scotland       f France        i Russia
3 Why is chicken no longer considered a luxury food today?
4 List the variety of forms in which chicken can be purchased.
5 Identify the factors that have resulted in an increase in the demand for chicken.
6 Discuss whether chicken is more nutritious than red meat.
7 Why is there much controversy about chicken production? What are your views on this issue?
8 Why should poultry be thoroughly thawed prior to cooking?
9 What happens to chicken when it is cooked?
10 How much will the following chickens weigh?
   a size 8         b size 10       c size 12

## ▶ Let's investigate

1 **Investigate** what cooking methods are used to prepare these chicken products:
   a KFC                  d roast chicken
   b warm chicken salad   e chicken cacciatore
   c chicken schnitzel    f chicken kebabs
2 How often have you consumed poultry in the past week? List the types consumed. Identify the different ways this meat was bought/packaged?
3 **Design** a new takeaway meal that has poultry as its main ingredient. Describe the ingredients in the meal. Draw and label this meal and give it a creative name.

## ▶ e-Poultry

www.inghams.com.au

### e-RIDDLE
Q: Why did the chicken cross the basketball court?
A: He heard the referee call fowl.

Go to the 'Product' section of Ingham's website. Research the various chicken portions. Then complete these tasks.
1 List eight different varieties of chicken portions.
2 Describe what a drumstick is.
3 Identify the portions that are offered with and without skin.
4 What is the name of the skinless eye fillet of the chicken breast?
5 Which part of the chicken does the thigh come from?
6 What is the difference between a thigh fillet and a thigh?

## ▶ Puzzled

### True or false

1 The alphabet is used to indicate the weight of poultry.
2 The skin of chicken contains a lot of fat.

MEAT, FISH, POULTRY, EGGS, NUTS AND LEGUMES

3  The chicken is said to have descended from the jungle fowls in China.
4  You should only store chicken for several days in the refrigerator.
5  The consumption of chicken in Australia is declining.
6  Deep-frying is a low-fat method of cooking chicken.
7  Chicken is not an excellent source of protein.
8  Chicken meat will turn white during the cooking process.
9  Chicken is eaten in many countries.
10 The drumsticks refer to the chicken breast.

### Missing words

Select the correct word/s from the box to complete the statements.

> cock-a-leekie   moisture   fat   free range   India   roasting
> grilling   KFC   frying   skin   drumsticks   breast   poultry   Russia   China   protein
> tori teriyaki   nutrition   battery farming   skinless   iron

1  Another name for a range of farmyard-bred birds is ☐ _ _ _ _ _ _.
2  The most tender part of a chicken is the _ ☐ _ _ _ _.
3  A time-consuming method of cooking involving the oven is _ ☐ _ _ _ _ _ _.
4  Dry meat results when too much of the _ _ _ _ ☐ _ _ _ is released during cooking.
5  A famous fast food outlet renowned for selling chicken is _ _ _.
6  The method of farming that allows chickens to roam around in small flocks is _ _ _ ☐ _ _ _ _ _.
7  The part of the chicken that contains the most fat is the _ _ ☐ _.
8  The country from which the chicken dish cotletki pojarski comes is _ _ _ _ _ _.
9  While chicken and red meat both supply _ _ _ ☐, red meat provides more of it.
10 A traditional Scottish chicken dish is _ _ _ _ _ _ _ _ _ _ _.

Now, by using the letters in the boxes, work out which nutrient coagulates during the cooking process of chicken. Chicken is an excellent source of this nutrient: _ _ _ _ _ _ _.

### Aesop's saying

Complete this famous saying that Aesop, an ancient Greek writer of fables, is said to have stated to a thoughtless milkmaid: 'Don't count your _____ before they hatch'. What does it mean?

### Poultry puzzle

Find each of the words from the box in the puzzle.

> chicken   duck   goose
> guinea fowl   partridge   pheasant
> pigeon   quail   turkey

| G | H | N | I | D | R | N | B | Y | P |
|---|---|---|---|---|---|---|---|---|---|
| Z | U | W | O | E | U | T | K | E | A |
| M | H | I | S | E | G | C | P | K | R |
| X | F | O | N | T | G | H | K | R | T |
| Z | O | G | V | E | E | I | G | U | R |
| G | I | X | E | A | A | S | P | T | I |
| P | M | Q | S | D | I | F | P | I | D |
| U | F | A | P | O | U | N | O | A | G |
| Q | N | E | K | C | I | H | C | W | E |
| T | Q | U | A | I | L | V | T | O | L |

ONE FOR THE BIRDS

# Let's Produce

## Chicken filo parcels (makes 4)

### Ingredients

150 grams sweet potato, peeled and cut into 2-centimetre cubes
$\frac{1}{4}$ cup olive oil
$\frac{1}{2}$ Spanish onion
150 grams chicken fillets, cut into 2-centimetre cubes
8 spinach leaves, washed and shredded
125 grams cottage cheese
3 shakes black pepper
12 sheets filo pastry
2 teaspoons olive oil, extra
2 teaspoons sesame seeds

### Method

1. Preheat the oven to 200°C.
2. Grease a baking tray.
3. Coat sweet potato with two teaspoons of oil.
4. Place sweet potato on the baking tray and roast in the oven for thirty minutes or until tender. Set aside to cool slightly. Leave oven on, preheating.
5. Meanwhile, heat two teaspoons of oil in a frypan and cook onion for two minutes.
6. Add chicken and cook for five minutes, stirring occasionally to ensure it is cooked through.
7. Transfer chicken and onion to a large bowl and cool slightly.
8. Add cooked sweet potato, spinach, cottage cheese and pepper. Mix well.
9. Divide into four portions.
10. Lay pastry on a clean bench area and cover with a clean, dry towel and then a damp tea towel to prevent the pastry from drying out.
11. Brush one sheet of pastry with oil.
12. Top with another two sheets of pastry, brushing each with oil.
13. Place one portion of chicken mixture on the narrow end of the pastry, leaving a 10-centimetre border at the end and on each side.
14. Fold the front end over the chicken mixture and then fold in the sides and roll up.
15. Repeat the process with the other three portions.
16. Place on a greased baking tray. The side of the pastry with the seam should be placed downside on the tray.
17. Brush each parcel lightly with oil and sprinkle with sesame seeds.
18. Bake until crisp and golden brown (about twenty to twenty-five minutes).

MEAT, FISH, POULTRY, EGGS, NUTS AND LEGUMES

## Butter masala chicken (serves 2)

### Method

1. Heat two teaspoons of oil in a frypan and add the chicken.
2. Stir-fry for four minutes, or until brown, and remove from pan.
3. Add butter and garam masala, paprika, chilli powder, coriander, cardamom and cinnamon and stir-fry for one minute until aromatic.
4. Add cooked chicken and stir.
5. Add tomato purée and sugar and simmer uncovered for fifteen minutes, stirring occasionally.
6. Stir in cream, yoghurt and lemon juice and cook for a further five minutes, or until sauce thickens slightly.
7. Serve with cooked basmati rice.

### Ingredients
- 1 tablespoon oil
- 200 grams chicken thigh fillets, sliced
- 30 grams butter
- 1 teaspoon garam masala
- 1 teaspoon coriander
- 1 teaspoon paprika
- 1/4 teaspoon chilli powder
- 1/4 teaspoon cardamom
- 1/4 teaspoon cinnamon
- 175 millilitres tomato puree
- 2 teaspoons sugar
- 1/4 cup thin cream
- 35 grams natural yoghurt
- 2 teaspoons lemon juice
- 1/2 cup basmati rice

## Hoisin drumsticks (serves 2)

### Method

1. Preheat oven to 200°C.
2. Make two 5-centimetre-long diagonal slits in the surface of each of the drumsticks.
3. Mix hoisin sauce, golden syrup and oil together and brush over the drumsticks.
4. Place the drumsticks on a plate, cover with plastic wrap and refrigerate for fifteen minutes to marinate.
5. Line a baking tray with non-stick paper and place drumsticks on lined tray.
6. Cook for thirty minutes or until golden brown and cooked through. Baste the drumsticks with marinade every ten minutes.
7. Meanwhile, cook the rice in a large saucepan.
8. Wash and dry lettuce leaves and place on serving plate.
9. Top with chicken.
10. Serve rice as an accompaniment.

### Ingredients
- 2 tablespoons hoisin sauce
- 1 tablespoon golden syrup
- 2 teaspoons peanut oil
- 4 chicken drumsticks
- 1/2 cup basmati rice
- 30 grams mixed lettuce leaves

### QUESTIONS

1. What does the term *basting* mean?
2. Describe the flavour of your cooked chicken drumsticks.
3. What is the purpose of a marinade?
4. **Investigate** other ingredients you could use as a marinade.

ONE FOR THE BIRDS

## Chicken burger (serves 2)

**Ingredients**
- 100 grams chicken mince
- 1 green shallot, thinly sliced
- 1 clove garlic
- 1 teaspoon lemon juice
- 2 teaspoons fresh coriander, chopped
- 1 teaspoon red curry paste
- $\frac{1}{4}$ cup breadcrumbs, made from 1-day-old bread
- 1 tablespoon egg, lightly whisked
- 2 teaspoons olive oil
- 4 slices Lebanese cucumber
- 15 grams bean sprouts
- 25 grams lettuce mix
- 2 pieces Turkish bread, split in half
- sweet chilli sauce

**Method**
1. Preheat the oven to 180°C.
2. Combine chicken mince, shallots, garlic, lemon juice, coriander, curry paste, breadcrumbs and egg in a large bowl. Use your hands to mix until it is evenly combined.
3. Shape the mixture into two patties.
4. Place patties onto a tray lined with non-stick paper, cover with plastic wrap and refrigerate for fifteen to twenty minutes or until firm.
5. In the meantime, slice the cucumber and wash the bean sprouts and lettuce mix.
6. Heat olive oil in a frypan and cook the chicken burgers for about seven minutes on each side or until cooked through.
7. Slice the Turkish bread in half and fill with the lettuce mix, cucumber slices, bean sprouts and chicken burger. Top with sweet chilli sauce.

### e-HINT

- Beat several eggs in a bowl for students to collect one tablespoon of lightly whisked egg.
- Supermarkets sell Turkish-style bake-at-home bread, which can be cut into four. To get a nice crispy crust, run under cold water for five seconds, shake and heat in the oven for five to ten minutes.

## QUESTIONS

1. Why is it recommended that the burgers sit in the refrigerator prior to cooking them?
2. **Investigate** other salad ingredients you could use.
3. **Investigate** other breads you could use instead of Turkish bread.
4. **Investigate** other types of mince you could use to make burgers. What ingredients would you use with them?
5. Can you think of a more creative name for your chicken burger?

## Warm chicken salad (serves 1)

**Ingredients**
- 75 grams chicken breast fillets
- 2 teaspoons tandoori paste
- 50 grams mixed lettuce leaves
- $\frac{1}{8}$ red capsicum
- 4 grams snow peas
- 1 tablespoon corn kernels
- $\frac{1}{4}$ carrot, cut into julienne strips

**Dressing**
- 2 tablespoons natural yoghurt
- $\frac{1}{2}$ teaspoon lemon or lime juice
- 1 teaspoon fresh basil, chopped

**Method**
1. Brush chicken fillets with tandoori paste. Cover with plastic wrap and place in the refrigerator for at least fifteen minutes.
2. Preheat griller and cook chicken fillets for ten to fifteen minutes until cooked, turning once during the cooking.
3. Cool and cut into strips.
4. Mix yoghurt, lemon, lime juice and basil in a small bowl.
5. Wash and dry lettuce leaves and arrange in another bowl.
6. Mix in capsicum, snow peas, corn and carrot.
7. Place chicken on salad and spoon yoghurt dressing on top.

# Chapter 17

# Egged On: Discover Eggs

**e-fact**
Most of the eggs people eat are from hens.

Did you know that Australia produces approximately 2.4 billion eggs per year and that the average Australian consumes about 140 eggs per year?

EGGED ON

Eggs are very versatile: they can form a substantial part of breakfast, lunch or dinner and can be prepared in many ways, such as hard boiled, scrambled, fried and poached. They also perform many functions:
- enriching food by providing extra nutrients
- binding or holding food together, like fish or chicken burgers
- thickening dishes, such as custards
- coating foods to protect them during cooking, such as veal schnitzel (which is coated with egg, flour and breadcrumbs)
- aerating food, like meringues, where the whipping of the eggs lightens the mixture

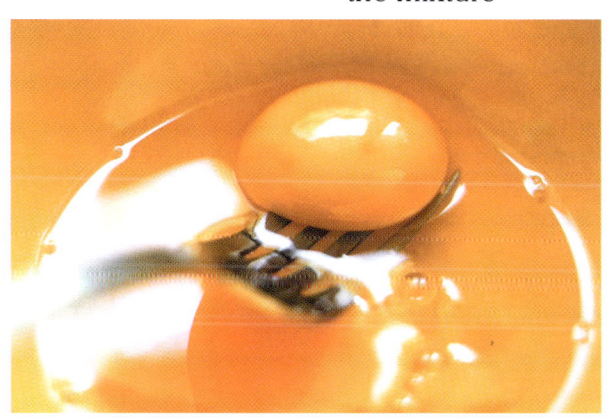

Eggs are associated with customs and celebrations in many different countries. For example, the Easter egg symbolises fertility and new life. So at Easter time many people give and receive chocolate eggs and, in some countries, people decorate them. In the Ukraine and Poland, eggs are decorated with simple colours and designs. In Germany, hollow eggshells are hung from dried branches, together with Easter decorations, similar to the way trees are decorated at Christmas. The eggs are hollowed by piercing the end of the eggshell with a needle and blowing out the contents.

# Classify Eggs

Eggs are classified according to either the type of bird that produces them, their size or their type, such as battery hen, vegetarian, omega-3, barn laid, free range or organic.

## Type of bird

Besides hens, other birds produce eggs that can be eaten, such as quails, ducks, geese and ostriches. Can you think of any other birds that produce edible eggs?

## Size of eggs

Eggs are sold in different sizes, which are determined by the egg farm that distributes them. When they arrive at the farm, a machine grades them by electronically weighing them. Some common sizes are 50, 55, 60, 65 and even 80 grams.

MEAT, FISH, POULTRY, EGGS, NUTS AND LEGUMES

## Types of eggs

| Egg type | Percentage of egg type produced | Description |
|---|---|---|
| Regular (battery hen) | 92.0 | Most eggs are laid by battery hens, which have their beaks clipped and are kept in small, crowded wire cages. |
| Vegetarian | 1.0 | Vegetarian eggs are produced by hens that have not been fed any animal products. |
| Omega-3 | 2.0 | Omega-3 eggs are produced from hens that have been fed omega-3 fatty acids and vitamin E. This enables the hens to produce eggs that are high in omega-3 fatty acids. |
| Barn laid | 2.5 | Hens that produce barn-laid eggs are kept in spacious barns with pens, where they are able to spread their wings and have more freedom. |
| Free range | 5.5 | Hens that produce free-range eggs are able to graze in open ground during the day. |
| Organic | 0.1 | Organic eggs are produced from hens that have been fed organically grown grain that is free from pesticides and fertilisers. To be sold as organic eggs, they must come from an organically certified organisation. |

Source: *Choice* magazine (www.choice.com.au/articles/a101882p1.htm)

## Properties of Eggs

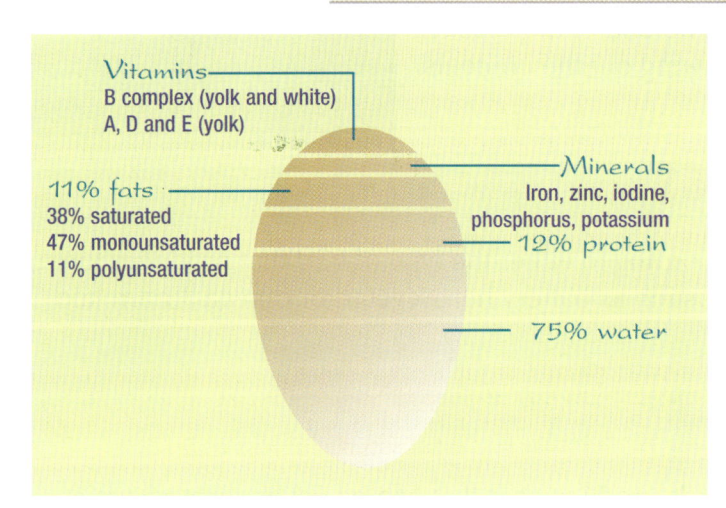

Vitamins
B complex (yolk and white)
A, D and E (yolk)

11% fats
38% saturated
47% monounsaturated
11% polyunsaturated

Minerals
Iron, zinc, iodine, phosphorus, potassium

12% protein

75% water

Eggs are an excellent source of protein (which is required by the body for the growth and repair of body tissue), low in fat (about 5 grams of fat, which is equivalent to one teaspoon of margarine) and contain a similar amount of cholesterol in the yolk. They are also a good source of vitamin $B_{12}$ and iron. The body is better able to absorb iron if vitamin C is present, so having a glass of orange juice when eating eggs is a good idea.

## Focus on Cholesterol

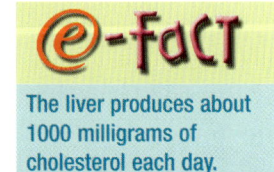

**e-fact**
The liver produces about 1000 milligrams of cholesterol each day.

Cholesterol is a fatty substance that the body produces. Some comes from food and has many good uses; however the body can make all that it needs. If we have too much cholesterol in our blood, this can cause fat to deposit in our blood vessels, thereby making it harder for the blood to flow through. In time this can cause the blood vessels to block and may lead to a heart attack.

EGGED ON

## WebExtras

**www.farmpride.com.au**
Farm Pride Foods is a grader, packer, processor, supplier and marketer of shell eggs and processed egg products within Australia.

**www.veggs.com.au**
Veggs Australia is one of the first companies in the world to market naturally enriched eggs that are low in saturated fats. The hens producing these eggs are fed a completely vegetarian diet.

**www.aeb.org**
The American Egg Board (AEB) is the US egg producers' link to consumers in communicating the importance of eggs.

It is now believed that the main causes of high blood cholesterol are eating too much saturated fat and being overweight. Your chances of having high blood cholesterol also increase if close family members have high levels. This means that, although eggs contain cholesterol and we once were told to limit our intake of eggs to one per week, it is fine to eat an egg four to five times per week.

## Egg Products

Most people use fresh eggs at home; however egg products make up a substantial part of the industry market. Egg products are eggs that have been processed by breaking open the shell and pasteurising the egg yolk and white. Those sold to industry include whole egg powder, egg white mix, cooked peeled eggs, pasteurised salted egg yolk and pasteurised scrambled egg in a bag. Next time you eat out and choose to have eggs, think about how they might have been processed prior to you eating them.

Some egg products are sold in supermarkets, such as Farm Pride Ready Whites, which are a mixture of pasteurised egg white, vegetable gum and a whipping agent. Using Ready Whites is a convenient and easy way to make meringues, pavlovas and soufflés. Can you think of two advantages and two disadvantages of using Ready Whites as opposed to fresh eggs?

### ▶ Let's remember

1. How many eggs does the average Australian consume?
2. List the uses of eggs. Provide one example for each.
3. What does the Easter egg symbolise?
4. How is the size of an egg determined?
5. What is a vegetarian egg?
6. What nutrients can be found in eggs?
7. How much fat is found in an egg?
8. In which part of the egg is cholesterol found?
9. Why is it a good idea to eat oranges and eggs at the same meal?
10. Describe an egg product and list one example.

### ▶ Let's investigate

1. Debate the following topic in class: 'People should only buy free-range eggs'.
2. You have recently been employed by the Royal Society for the Prevention of Cruelty to Animals (RSPCA) as a campaign project officer. Your first task is to oversee the 'Hens out of cages' campaign, which urges the public to 'think before choosing their eggs'. You have been asked to investigate the information on the website www.rspca.org.au and to write a letter to your State premier, outlining the position of the RSPCA.

MEAT, FISH, POULTRY, EGGS, NUTS AND LEGUMES

3 Choose a country and `investigate` the use of eggs in the celebrations of that country.
4 Barn-laid and free-range eggs are about $1 more per dozen than regular eggs. Is this true? `Investigate` the price of the different types and sizes of eggs. Construct a table to show your results.
5 Using the article below as a stimulus, `investigate` some of the factors that might increase the number of eggs hens lay.

## I could be so clucky: Kylie helps hens lay eggs

A Shropshire poultry farmer claims playing Kylie Minogue songs to his hens makes them lay more eggs.

Ray Link found his hens laid more eggs after he played music, but discovered their weekly haul was even bigger when he put on Kylie's hits.

Mr Link, who runs Gobbets Rare Breeds Poultry farm, near Ludlow, heard about cows' milk yield being boosted by music so he tried music on his hens.

He said: 'I'd nothing to lose, so I put my wife Margaret's tape player out with the birds.

'Whenever Kylie is on they start laying. They love her so much I've bought her one of her albums.'

The *Sun* reports that Ray has also thought about showing his hens a video of Kylie in *Neighbours*.

Source: www.ananova.com

### e-Eggs

**www.goldeneggs.com.au**

Search in the 'Farm to Family' section and draw a flow chart to show what happens to eggs after they are laid and before consumers purchase them. Include an explanation of the terms *haugh measurement*, *grading*, *stamping* and *candling*.

**www.choice.com.au/articles/a101881p1.htm**

*Choice* magazine answers some frequently asked questions about eggs. Summarise the information about eggs using these headings:
- storage
- shelf-life
- freshness

**www.canadaegg.ca/english/educat/teach.html**

The Canada Egg Board website has an 'Extraordinary Egg Series' of teaching resources that consists of five teaching modules. Each module contains activity suggestions, activity master sheets, teacher notes and additional resources. A fabulous website for teachers, though requiring some modification for use in Australia.

### e-fact
Want to know if an egg is fresh? Place it in a bowl of water and if it is stale, it will float.

### Puzzled

#### Let's get cooking
In New Jersey, USA, there is a town called Egg Harbour. If you were a town planner and were able to name a city and its streets after eggs, what would you choose and why?

#### Mathematical egg-xample
If a hen takes twenty-four hours to produce an egg and then starts producing again after thirty minutes, how many eggs will she produce in one month?

EGGED ON

# Let's Produce

## French toast (serves 2)

**Ingredients**
- 1 banana
- 1 kiwi fruit
- 6 strawberries
- 1 egg
- ⅓ cup milk
- 4 slices thick bread
- 30 grams butter
- icing sugar, for dusting
- ½ cup yoghurt
- 2 tablespoons maple syrup

### Method

1. Peel and slice the banana and kiwi fruit.
2. Hull the strawberries—that is, remove the stalk—and slice in half.
3. Combine the egg and milk.
4. Soak the bread in the egg and milk mixture for one to two minutes.
5. Melt butter in a frypan and fry bread on both sides until golden brown.
6. Serve immediately, dusting with icing sugar. Top with fresh fruit, yoghurt and maple syrup.

### Variations

Serve with either grilled bacon and banana, scrambled eggs or your own selection of fruit, such as apricots, mango, pineapple, pear or peaches.

## Spinach and feta pie (serves 4)

**Ingredients**
- 6 sheets filo pastry
- ½ bunch silverbeet
- 1 onion, finely chopped
- 125 grams feta cheese
- 60 grams Parmesan cheese
- 4 eggs, lightly beaten
- pepper, to taste

### Method

1. Brush three filo sheets with butter and line the base of a greased pie dish.
2. Wash and drain chopped silverbeet. Add onion.
3. Mix the silverbeet with feta cheese, Parmesan cheese, eggs and pepper.
4. Press silverbeet mixture into dish. Brush remaining three sheets of filo pastry. Cover mixture with filo pastry.
5. Bake in a moderate oven for thirty to forty minutes.

MEAT, FISH, POULTRY, EGGS, NUTS AND LEGUMES

### Ingredients
- 60 grams butter
- 100 grams cooking chocolate
- 2 large eggs
- 2 tablespoons plain flour
- $\frac{1}{3}$ cup caster sugar

## Chocolate soufflé pudding (serves 2)

### Method
1. Grease two soufflé or ramekin dishes with butter.
2. Melt butter and chocolate together in microwave on medium high for one to two minutes or until just melted.
3. Separate egg yolks from egg whites.
4. Stir yolks into cooled butter and chocolate mixture. Stir in flour.
5. Beat egg whites until soft peaks form. Add caster sugar and beat until peaks are firm.
6. Fold egg white mixture into flour mixture.
7. Place mixture into dishes and bake for thirty to forty minutes.
8. Serve with ice cream.

### Ingredients
- 125 grams fettuccine
- pinch salt
- 1 tablespoon oil
- 2 rashers bacon
- $\frac{1}{4}$ cup cream
- 2 eggs
- 30 grams Parmesan cheese
- freshly ground pepper

## Fettuccine carbonara (serves 2)

### Method
1. Place fettuccine into a large saucepan of boiling water. Add a little salt for flavour and one tablespoon oil to prevent the pasta from sticking.
2. Cut bacon into thin strips. Cook bacon in a frypan over a low heat until just crisp.
3. Add cream to pan. Add lightly beaten eggs and half of the Parmesan cheese. Cook over heat very gently until mixture has thickened and has heated through.
4. Stir in cooked fettuccine.
5. Serve immediately with freshly ground pepper and remainder of Parmesan cheese.

### QUESTIONS
1. What would happen if the cream and egg mixture was overcooked or overheated?
2. Suggest some other ways to cook bacon other than in the frypan.
3. What are some of the ways you can purchase Parmesan cheese?

# Potato and ham frittata (serves 4)

**Ingredients**
1 tablespoon extra virgin olive oil
½ leek, finely sliced
1 clove garlic, crushed
2 medium potatoes, cooked and cut into 0.5-centimetre-thick slices
4 slices ham, diced
3 tablespoons grated Parmesan cheese
3 eggs
salt and pepper

## Method

1. Heat oil in a shallow frypan. Sauté the leek and garlic until transparent. Add the potatoes and ham and combine gently.
2. In a small bowl combine the Parmesan cheese, lightly beaten eggs and seasoning. Pour egg mixture over the potato mixture and cook on a low heat until almost set.
3. Place the pan under a preheated grill and cook until light, golden brown and egg mixture appears puffy.

## QUESTIONS

1. What is the meaning of the term *sauté*?
2. Why is it important to cook the egg mixture over a low heat?
3. The term *seasoning* is used in this recipe. What does it mean?

EGGED ON

# Chapter 18

# Going Nuts: Discover Nuts

**e-fact**

Did you know that when astronauts fly into space, the only food they do not have to treat and dehydrate are pecan nuts?

Nuts are plant foods: they are the edible dried fruits or seeds of plants, have a hard shell and tend to grow on trees. One exception is peanuts because they are not a true nut: they are really legumes and grow under ground. However they are grouped with nuts as they share many of their properties.

We should eat about 30 grams of nuts five times a week. Children aged under five are not advised to consume nuts because they may not chew them properly and therefore choke. We should all ensure that nuts are a regular part of our diet because they:
- are very nutritious
- assist with reducing the risk of heart attack and coronary heart disease
- form part of a varied, plant-based diet

**e-define**

Rancid means that a food, in this case nuts, develops an unpleasant and 'sour' taste.

Nuts can become rancid because of their high fat content. It is therefore important to store them correctly. Nuts will keep much longer when stored in their shells; if removed from their shell, they should be kept in an airtight container in a cool place. A refrigerator is an ideal place to store them.

GOING NUTS

pistachio nuts • peanuts • brazil nuts • chestnut • coconut • walnuts • hazelnuts • almonds • macadamia nuts • pecan nuts • cashew nuts

Nuts are often roasted to increase their flavour and make them crunchier. However you must be careful because they can burn quite easily due to their high fat content.

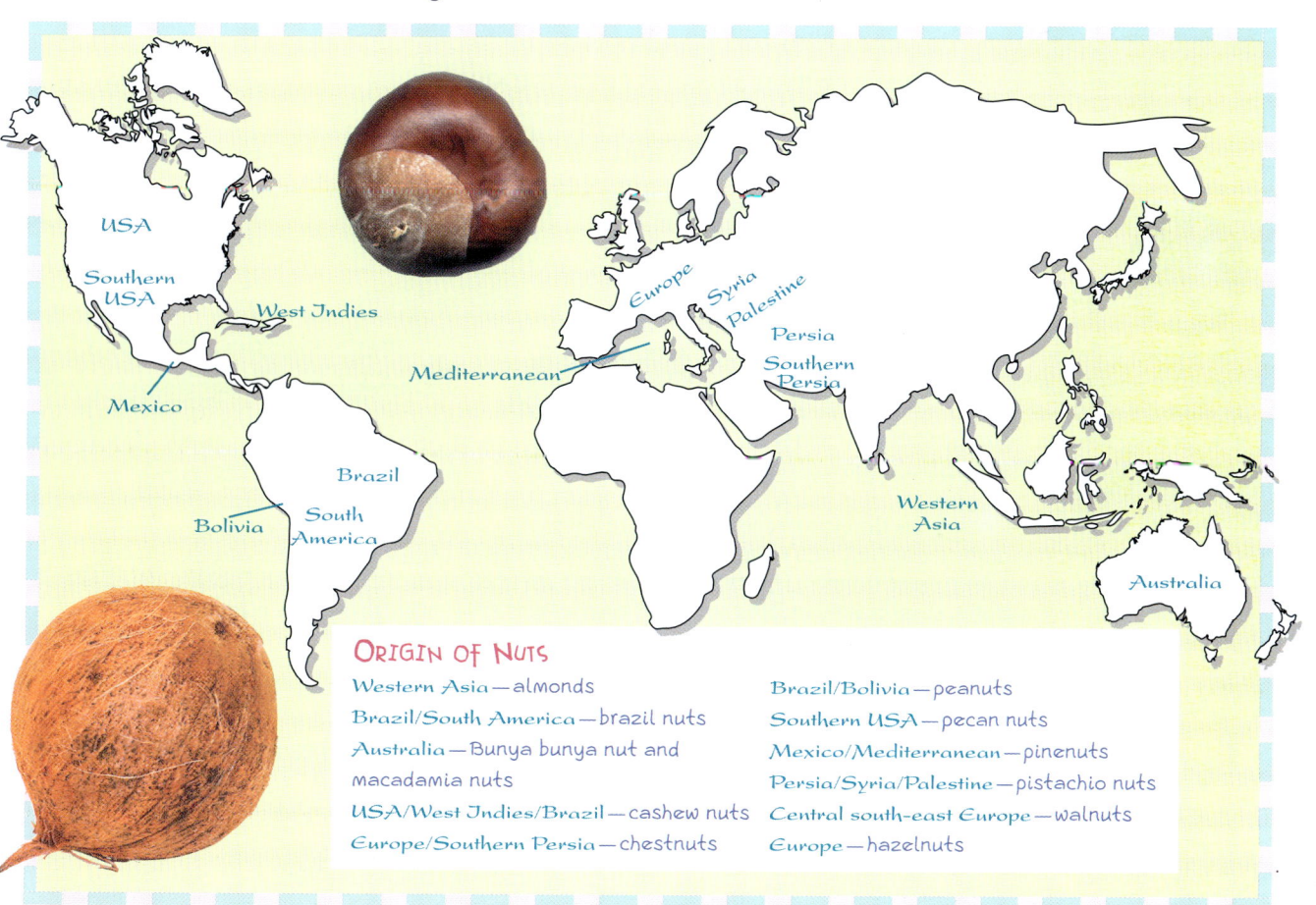

### Origin of Nuts

Western Asia — almonds
Brazil/South America — brazil nuts
Australia — Bunya bunya nut and macadamia nuts
USA/West Indies/Brazil — cashew nuts
Europe/Southern Persia — chestnuts
Brazil/Bolivia — peanuts
Southern USA — pecan nuts
Mexico/Mediterranean — pinenuts
Persia/Syria/Palestine — pistachio nuts
Central south-east Europe — walnuts
Europe — hazelnuts

MEAT, FISH, POULTRY, EGGS, NUTS AND LEGUMES

# Properties of Nuts

**e-fact**
You can roast nuts in the microwave by placing them in an oven bag and twisting the opening to secure it. Cook on high for about five minutes, gently shaking the bag after each minute.

Nuts are quite nutritious. Many people in the past have avoided eating them because of their high fat content; however many health benefits can be gained from regularly eating a small amount. The fats present in nuts are regarded as 'good' because they are mainly monounsaturated and polyunsaturated. The only exception are fats present in coconuts and palm nuts: they tend to be saturated and may contribute to heart disease. Being plant foods, nuts contain no cholesterol. We often use the fats extracted from them in cooking. Where do you think walnut oil comes from?

Nuts are also high in protein, carbohydrates and fibre; the protein is incomplete, which means that some of the essential amino acids are missing. Nuts play an important role in the diet of vegetarians, so they should be eaten with other sources of protein, such as legumes or vegetables, to ensure that all of the amino acids are present.

**e-fact**
Did you know that 3 kilograms of peanuts makes 1 litre of peanut oil? Now, that's a lot of peanuts!

# Focus on Incomplete Protein

**e-fact**
Blanching nuts means their skins are removed.

As mentioned above, the protein present in nuts is incomplete: it does not contain all of the eight essential amino acids. Vegetarians can balance their intake of amino acids by eating the following combinations of food:
- legumes with seeds and nuts
- legumes with grains and cereal products
- vegetables with seeds and nuts
- vegetables with grains and cereal products

## Case Study: Peanut Butter

The grinding of peanuts to make an edible paste occurred over 1000 years ago in South America; however the manufacture of the peanut butter with which we are familiar today began in the United States over 100 years ago. Peanut butter was first manufactured in Australia by Sanitarium in 1898.

Peanuts grow in the ground and are therefore cleaned of dirt before being shelled and roasted. They are then blanched, such as by brushing them with bristles or rolling them against rubber rollers. The nuts are next passed through a light beam to ensure they are whole, roasted and blanched. They are ground into either smooth or crunchy pastes; some crunchy varieties have finely chopped nuts added after grinding. The peanut butter is cooled before being poured into jars. It is then distributed to supermarkets for the consumer to buy.

### Questions

1. Explain the process of making peanut butter by creating a flow chart. The following headings should be used:
   - cleaning and preparation
   - quality control
   - distribution
   - roasting
   - grinding
   - blanching
   - packing
2. How would you describe the appearance, flavour, texture and aroma of peanut butter?
3. What is your favourite type of peanut butter? Why?

GOING NUTS

## ▶ Let's remember

1. Are peanuts really a nut? Explain your answer.
2. Why should nuts be a regular part of your diet?
3. Explain why it is not recommended for children aged under five to consume nuts.
4. How should nuts be stored?
5. What does the term *rancid* mean and why are nuts subject to rancidity?
6. Why are nuts often roasted?
7. Why should safety considerations be observed when roasting nuts?
8. Which nuts originated in Australia?
9. Why are the fats present in most nuts considered to be good?
10. In what ways can nuts be included in the diet of a vegetarian to ensure that all of the amino acids are present? Provide specific examples of menu selections.

## ▶ Let's investigate

1. **Investigate** the various ways nuts can be purchased by visiting your local supermarket, browsing through your pantry at home or school or visiting a website, such as www.colesonline.com.au or www.woolworths.com.au Complete a table like the one below for the different nuts. Almonds have been completed for you.

   | Nut     | Varieties                          |
   |---------|------------------------------------|
   | Almonds | Kernels, flaked, ground, slivered  |

2. Graph the fat and protein percentages present in a variety of nuts by using the data provided in the table below.

   | Nut           | Fat (%) | Protein (%) |
   |---------------|---------|-------------|
   | Almonds       | 52      | 20          |
   | Walnuts       | 62      | 15          |
   | Cashew nuts   | 46      | 15          |
   | Pistachio nuts| 48      | 22          |
   | Brazil nuts   | 66      | 14          |
   | Peanuts       | 49      | 26          |

   a. Which nut/s have the:
      i. highest percentage of fats
      ii. lowest percentage of fats
      iii. highest percentage of protein
      iv. lowest percentage of protein
   b. Which nut/s do you think are the healthiest? Explain your answer.
   c. Why do you think you have to be very careful when roasting nuts? Link your discussion to the percentage of fat present in most nuts.

3. Draw your five favourite nuts in your workbook.

4. Using a range of recipe books or websites, find ten recipes that include nuts.
   a. What types of nuts are used?
   b. Are the recipes for savoury or sweet food?
   c. How are the nuts prepared—whole, crushed or roasted, for example?
   d. Present your information to the class.

---

### WEB EXTRAS

**www.karsnuts.com**
The Kar Nut Products Company provides nut and snack items throughout Michigan, USA. Its website gives information on not only its product range but also the history of nuts, their nutritional information and lots of handy links.

**www.agrimac.com.au**
Agrimac Macadamias is an Australian-owned plantation manager, processor, marketer and consultant on macadamia nuts. Its website provides a brief history of macadamia nuts, their nutritional information and a few delicious recipes.

**www.sanitarium.com.au/schoproj/pppeanut.htm**
This Sanitarium webpage entitled 'The Story of Peanut Butter' provides detailed project information.

**www.almondco.com.au**
Almondco Australia has been processing almonds for over half a century. Besides its product range, this website provides information on the Australian almond industry, their nutritional value and recommended storage conditions.

MEAT, FISH, POULTRY, EGGS, NUTS AND LEGUMES

## e-Nuts

www.macadamias.org

### e-RIDDLE

Q: Why was the nut too scared to walk through the park?
A: He was afraid he might be a-salt-ed.

1 Click on the 'General info' link.
   a What did the Aboriginal people call macadamia nuts?
   b Who was the macadamia nut named after?
   c If the macadamia nut was named after you, what would it be called?
   d Draw a map of Australia and colour in where macadamia nuts are commonly grown.
   e Describe the trees that macadamia nuts grow on.
   f What does the term *dehusking* mean?
   g Why is drying an important part of the processing of macadamia nuts?
   h Describe how macadamia nuts are now cracked.
2 Click on 'Health benefits'.
   a Describe the type of fat present in macadamia nuts.
   b Outline the health benefits of consuming macadamia nuts.
   c In what ways can you consume macadamia nuts?
   d Graph the nutritional information of macadamia nuts. To better present your work, you might like to use a computer spreadsheet program.

## Puzzled

### Brainteaser

Why do you think the ancient Greeks and Romans believed that walnuts could cure headaches? Think about the shape of the walnut.

### True or false

Copy these statements into your workbook and indicate whether they are true or false.
1 The peanut is not a true nut.
2 Nuts are generally high in fat.
3 Nuts are generally very high in cholesterol.
4 All nuts grow under ground.
5 The cashew is a round nut.
6 The best place to keep nuts is in an airtight container in a cool place.
7 The coconut is a large nut.
8 Nuts are generally high in folate.
9 The coconut contains a lot of saturated fats.
10 Peanut butter is made from grinding peanuts.

### Nut scramble

Unscramble the following ways nuts can be purchased.
1 stadeot
2 kedelf
3 leowh
4 deshruc
5 popdech
6 valehd

GOING NUTS

## Nutting out nuts

Using the photographs and descriptions, identify the following nuts.

| Name of nut | Description | Nut |
|---|---|---|
|  | • Long, flat, oval shaped<br>• Ridged surface |  |
|  | • Large and long<br>• Tough<br>• Brown–grey colour |  |
|  | • Creamy brown colour<br>• Round shell, with two hemispheres shaped like 'butterflies' |  |
|  | • Small green kernel<br>• Thin yellow–red skin<br>• Smooth, brittle, creamy coloured shell |  |
|  | • Round<br>• Very smooth<br>• Hard, smooth brown shell |  |
|  | • Very large<br>• Fruit of a palm tree<br>• Brown fibrous husk<br>• Brown shell<br>• Soft white flesh |  |
|  | • Large and round<br>• Greyish white<br>• Thin brown skin<br>• Smooth, soft brown shell |  |
|  | • Cream coloured<br>• Kidney shaped |  |
|  | • Creamy white in colour<br>• Oval shaped, flat with pointed ends<br>• Brown skin<br>• Light brown, flat, pitted shell |  |
|  | • Round and small with a point at the top<br>• Thin, brown skin<br>• Smooth, hard shell with a rough area at the base |  |

179

MEAT, FISH, POULTRY, EGGS, NUTS AND LEGUMES

# Let's Produce

## Choc nut cookies (makes 20)

**Ingredients**
- 60 grams butter
- ½ cup brown sugar
- ½ egg, lightly beaten
- ½ cup plain flour
- ½ teaspoon bicarbonate of soda
- 60 grams choc chips
- 30 grams peanuts

**Method**
1. Preheat the oven to 180°C.
2. Line a baking tray with non-stick baking paper.
3. Cream butter and sugar.
4. Add the egg and beat until combined.
5. Sift flour and bicarbonate of soda and add to the butter mixture.
6. Stir in peanuts and choc chips with a large metal spoon.
7. Place tablespoonfuls of mixture on the lined trays, leaving space for spreading.
8. Flatten each biscuit with a knife.
9. Bake for fifteen to twenty minutes or until light golden.
10. Let the biscuits cool on the trays for five minutes before transferring to wire racks to cool.

## Fresh fruit and nut pastries (serves 2)

**Ingredients**
- 1 small banana
- 1 ripe peach
- 1 lemon, juiced
- 1 tablespoon muscatel raisins
- 2 tablespoons apple juice
- 2 teaspoons maple syrup
- 4 sheets filo pastry
- 1 tablespoon almond meal
- 2 teaspoons butter, melted
- ½ cup vanilla yoghurt

**Method**
1. Preheat oven to 180ºC.
2. Line a baking tray with non-stick baking paper.
3. Peel banana and cut diagonally into four pieces.
4. Cut peach in half and remove stone; slice each half into three pieces.
5. Drizzle peach and banana with lemon juice.
6. In a small saucepan, place raisins, apple juice and maple syrup and bring to the boil.
7. Simmer for five minutes over a medium-low heat until the juice has a thick, syrupy consistency.
8. Fold the filo pastry in half and place on a chopping board.
9. Cut out a circle shape approximately 15 centimetres in diameter.
10. Place two circles of single filo pastry on the baking tray and sprinkle each with one teaspoon of almond meal.
11. Brush the pastry edges with butter.
12. Roll about 3 centimetres of the edge of each filo circle towards the centre three times. Press firmly to create a border.
13. Arrange banana and peaches on each filo circle.
14. Spoon the raisin syrup over each pastry.
15. Bake for fifteen to twenty minutes or until golden.
16. Serve warm with vanilla yoghurt.

GOING NUTS

## QUESTIONS

1. What is the purpose of drizzling lemon juice over the banana and peach?
2. Why do you need to work quickly with the filo pastry sheets?
3. Where does almond meal come from?
4. What other nuts could you use?
5. What other fruit could you use?

### Ingredients

¼ cup roasted salted peanuts
3 tablespoons oil
water crackers

## Peanut butter

### Method

1. Place peanuts and 1½ tablespoons of oil in a food processor and blend until smooth.
2. Gradually add the remaining oil—if required for the correct consistency.
3. Serve with water crackers.

### Ingredients

150 grams dried flat rice stick noodles
¼ cup roasted salted nuts, chopped
1 teaspoon chilli paste
200 grams chicken thigh fillets, sliced into 2-centimetre strips
1 tablespoon peanut oil
100 grams snake beans, trimmed and sliced diagonally
1 carrot, peeled and cut into matchsticks
1 clove garlic, crushed
2-centimetre pieces of ginger, grated
2 tablespoons soy sauce
2 teaspoons lime juice

## Chilli chicken with nuts (serves 2)

### Method

1. Soak noodles in warm water until soft (about ten minutes). Drain.
2. Combine peanuts and chilli paste and add the chicken.
3. Heat one teaspoon of oil in a frypan and add half the chicken. Stir-fry until just cooked; transfer to a plate and repeat with another teaspoon of oil and remainder of chicken.
4. Add remainder of oil and stir-fry beans, carrots, ginger and garlic for two minutes.
5. Add chicken, noodles, soy sauce and lime juice and combine.
6. Serve immediately.

# Chapter 19

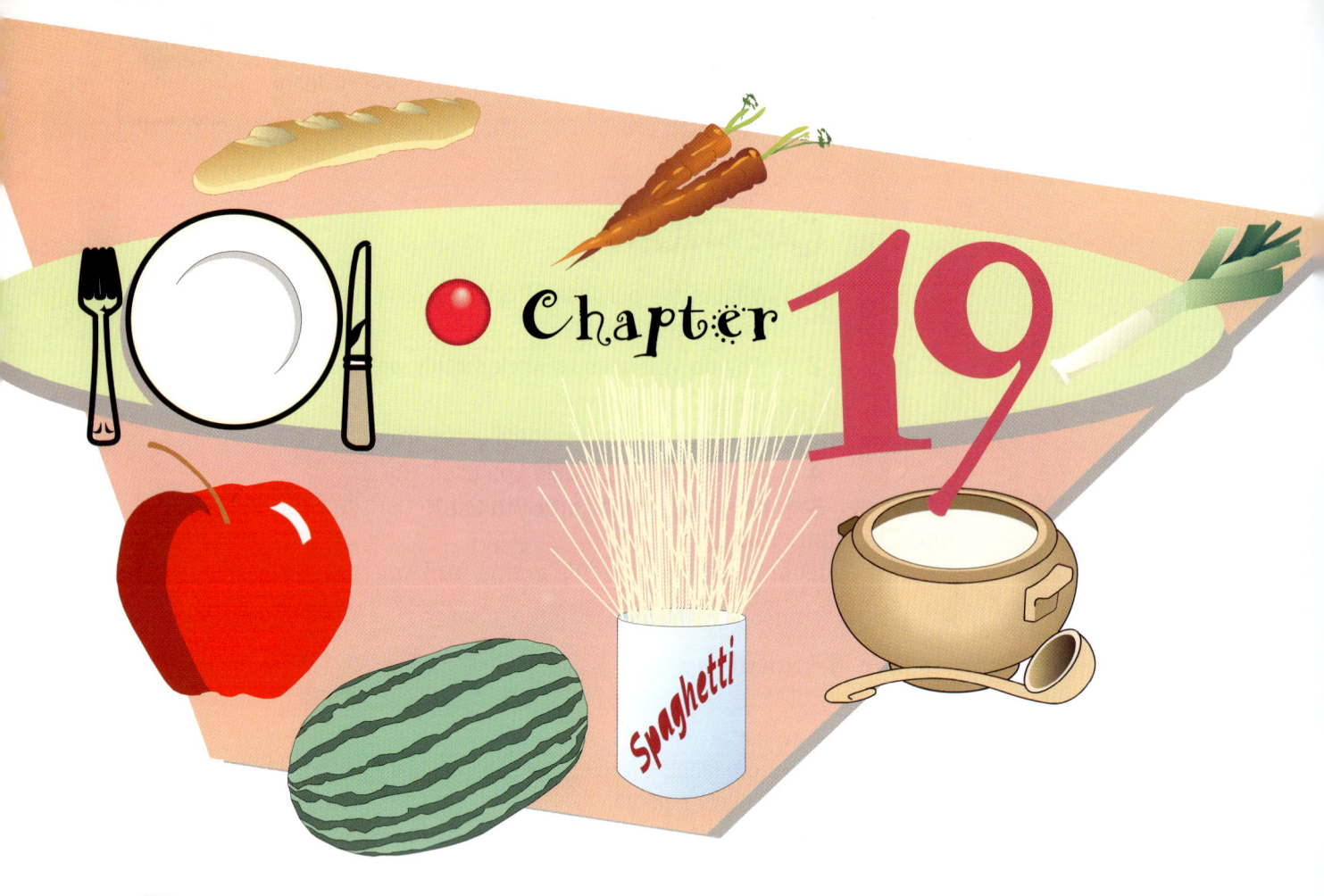

# Pulses Rising: Discover Vegetarians

Vegetarians do not eat meat; they eat foods from plants, such as grains, cereals, nuts, vegetables, fruits and legumes. Some vegetarians consume the products of animals, such as eggs, milk and milk products, while others do not.

## Classify Vegetarians

There are many different types of vegetarians, including:
- Lacto-ovo-vegetarians do not eat meat but include eggs, milk and milk products in their diet.
- Lacto-vegetarians do not eat meat or eggs but include milk and milk products in their diet.
- Ovo-vegetarians do not eat meat, milk or milk products but do consume eggs.

> **e-fact**
>
> A hint to help you remember the different types of vegetarians:
> - Lacto refers to milk.
> - Ovo refers to eggs.
> - Pollo refers to chicken.

- Vegans do not eat meat or the products of animals, such as eggs, milk and milk products.
- Pollo-vegetarians eat chicken, fish, eggs and milk, along with plant foods.

## Why Do People Become Vegetarian?

Some reasons why people become vegetarian include:
- Religion: Historically, vegetarianism is linked to certain Eastern religions and Christian groups.
- Health: A diet consisting of cereals, grains, fruit, vegetables, legumes and nuts is considered healthier than one that contains meat and meat products. This is one of the main reasons why vegetarianism is becoming more popular in Australia today.
- Animal welfare: There are people who believe it is wrong to kill animals for food or may object to the way the animals are raised/kept in intensive farming.
- Environmental issues: The production of meat is expensive compared with that of cereals and other similar crops—they consider it to be a wasteful resource that pollutes the environment.
- Taste: Some people do not like the taste or texture of meat.

About 20 per cent of the Australian population consume a mainly vegetarian diet. It is estimated today that at least 45 per cent have at least one vegetarian diet per week, and that about 20 per cent restrict their meat intake but still eat chicken and fish. The reasons for the increase in vegetarianism may be:
- the greater availability of interesting vegetarian dishes
- concerns over animal welfare
- the migration of people from cultures with a vegetarian trend

While a significant number of people do consume less meat in their diet, the actual number of dedicated vegetarians in Australia is fairly consistent at about 5 per cent of the population.

Studies show that vegetarians tend to have a lower risk of developing such diseases as coronary heart disease, high blood pressure, diabetes, obesity and some cancers. This may be because their diet consists of lots of fruit and vegetables, which contribute bulk, and generally has lower saturated fats. However the differences in health cannot solely be explained by the lack of meat in their diet. Other factors, such as control of alcohol and tobacco intake and greater physical exercise, may contribute to their greater health.

> **e-fact**
>
> Some famous vegetarians are:
> - Pythagoras
> - Socrates
> - Aristotle
> - Plato
> - Tolstoy
> - Brian Adams
> - Kim Basinger
> - Melissa Etheridge
> - Elton John
> - Brooke Shields
> - Anthony Robbins
> - Julian Lennon
> - Anthony Hopkins
> - Tom Cruise
> - Penélope Cruz
> - Bob Geldof
> - David Duchovny
> - Cher
> - Whoopi Goldberg
> - Ashley Judd
> - Michael J. Fox
> - Guy Pearce
> - Peter Brock
> - Ricki Lake
> - Jerry Seinfeld
>
> Do you know anyone who is a vegetarian? What type of vegetarian are they?

MEAT, FISH, POULTRY, EGGS, NUTS AND LEGUMES

# Focus on Vegetarian Diets

There are still nutritional concerns with a vegetarian diet. The more restrictive a vegetarian's diet, the greater the care that needs to be taken to ensure an adequate intake of all nutrients. Some people, especially teenage girls and young women, not only do not eat meat but also do not ensure they are obtaining a balance of nutrients. They should be looking for alternative food that would supply a similar relationship of nutrients to meat. Vegans, who do not consume milk, may lack calcium in their diets.

Look at the table below to see how a vegetarian can supplement his or her diet to assist with gaining all of the necessary nutrients.

| Nutrients that vegetarians may lack | Function | Vegetarian supplements | Additional comments |
|---|---|---|---|
| Protein | Growth of all body tissues | • Legumes, such as chickpeas and lentils<br>• Grains/cereal products, such as brown rice and couscous<br>• Vegetables, such as beans and peas<br>• Nuts<br>• Milk and milk products | Protein consists of amino acids. Unfortunately, protein from vegetable sources tends to be deficient in at least one of the essential amino acids. Vegetarians who eat a wide variety of food can overcome this problem as the essential amino acids can be balanced by combining grains with cereals or nuts with legumes. |
| Iron | To carry oxygen in the bloodstream | • Green leafy vegetables, such as spinach<br>• Legumes, such as baked beans<br>• Wholegrain cereals/grains<br>• Seeds, such as pumpkin and sunflower seeds<br>• Fruit, such as dried apricots, pears, peaches and raisins<br>• Tofu | One of the greatest nutritional problems for vegetarians is obtaining enough iron to prevent tiredness or fatigue and anaemia. Unfortunately, the iron in some vegetables is not readily absorbed. Vegetarians need to ensure they eat vitamin C-rich food with other food that contains iron to assist with the absorption of iron, such as oranges with breakfast cereal or tomatoes and capsicum with legumes. Also, it is useful not to drink tea or coffee with meals as this interferes with the absorption of iron. This is because whole grains have substances called phytates in their bran and the vegetables have a substance called oxalic acid, both of which prevent the iron from being absorbed in the bloodstream. |
| Zinc | To assist with the absorption of other nutrients | • Wholegrain cereals/grains<br>• Legumes, such as soybeans<br>• Nuts<br>• Seeds, such as sesame and sunflower seeds | The substances found in bran (phytates) and soybeans and soy milk (phosphates) can interfere with the amount of zinc absorbed. |
| Vitamin $B_{12}$ | Required for healthy nerve tissues | • Cheese, yoghurt, milk<br>• Eggs<br>• Tofu<br>• Mushrooms | All vitamins are supplied by a vegetarian diet, except vitamin $B_{12}$. This is present in animal products, so it is not a concern for those vegetarians who consume them. However a vegan has to ensure that they consume $B_{12}$ supplements. Soy milk may be fortified with $B_{12}$, meaning that this vitamin is added to the milk. |
| Calcium | To harden (or ossify) bones and teeth | • Fortified soy milk<br>• Tofu<br>• Nuts | A lack of calcium can contribute to osteoporosis. |

PULSES RISING

## ▶ Let's remember

1. List the different types of vegetarians and explain what food they can eat.
2. Provide reasons why people decide to become vegetarian.
3. Identify the trends associated with vegetarianism in Australia today.
4. Outline the dietary-related diseases that are less prevalent for vegetarians.
5. What reasons are given for a vegetarian being considered to be healthier than a non-vegetarian?
6. How can vegetarians ensure they consume all of the essential amino acids?
7. Why is it difficult for a vegetarian to obtain an adequate supply of iron?
8. How can a vegetarian improve the absorption of iron in their diet?
9. List the types of food from which a vegan can obtain calcium.
10. What vitamins are missing from a vegan's diet and how can they be consumed to ensure an adequate intake?

## ▶ Let's investigate

1. Copy the table below into your workbook. From the variety of food listed below, indicate which are suitable for the different types of vegetarians. Some may be eaten by more than one type of vegetarian.

peanuts, carrots, fish, cream, bagels, tomatoes, beetroot, yoghurt, tofu, beef, goat's milk, almonds, parsnips, bok choy, apples, cheese, soy milk, chickpeas, veal, eggs, cream cheese, butter, margarine, olive oil, lima beans, cabbage, lentils, steak, flavoured milk, chicken, oysters, sour dough bread, ham, wholemeal crumpets, onions, capsicum, lemons, kidney beans, hummus, bananas, rabbit, buttermilk, macadamia nuts, pasta, goose, free-range eggs, pears, asparagus, liver

| Lacto-ovo-vegetarian | Lacto-vegetarian | Vegan |
|---|---|---|
|  |  |  |

2. **Investigate** the consequences, including possible dietary-related diseases, for a vegetarian who does not consume adequate amounts of the following nutrients:
   - calcium
   - iron
   - vitamin $B_{12}$
   - zinc
   - complete proteins
3. Find out the names of the eight essential amino acids that adults need to consume in their diet. What is the name of the ninth essential amino acid that children need to consume?
4. **Design** a menu for a vegan that would incorporate all of the necessary nutrients. Justify your choice of food and nutrients. Include meal suggestions for breakfast, lunch and dinner, as well as snacks.
5. **Investigate** which religious groups have a vegetarian diet.
6. Select two recipes that contain meat and adapt them so that they would be suitable for a vegetarian to consume.
7. Animal and seafood are an excellent source of zinc:
   - 150 grams of grilled rump steak provides 7.5 milligrams.

## WEB EXTRAS

**www.vegetarian.com.au**
Vegetarian.com Australia serves as a gateway to everything you ever wanted to know about vegetarianism, providing links to products, services, health, relaxation, education and environment.

**www.vnv.org.au**
Vegetarian Network Victoria (VNV) is a non-profit, volunteer organisation whose aim is to promote the benefits of a vegetarian lifestyle and to provide vegetarians with service, information and support.

**www.moreinfo.com.au/avs/vegdef.html**
The Australian Vegetarian Association website provides instructive information on the health, economic and environmental benefits of vegetarianism.

MEAT, FISH, POULTRY, EGGS, NUTS AND LEGUMES

- Twelve oysters supplies 65 milligrams.
- 150 grams of grilled fish provides 1.4 milligrams.

**a** The plant-based food listed in the table below serves as a source of zinc for vegetarians. Create a graph to illustrate the information.

| Plant food | Amount of zinc (mg) |
| --- | --- |
| 2 slices wholemeal bread | 0.7 |
| 1 cup cooked rolled oats | 0.5 |
| 2 Weet-Bix | 0.4 |
| 1 cup canned baked beans | 0.8 |
| 1 cup cooked lentils | 1.4 |
| 1 cup cooked soybeans | 2.6 |
| 2 cups cooked wholemeal pasta | 1.8 |
| 1 tablespoon peanut butter | 0.7 |
| 50 g roasted cashews | 2.7 |
| 50 g almonds | 1.7 |
| 50 g pumpkin seeds | 3.3 |

**b** It is recommended that adults and children over twelve years consume 12 milligrams of zinc per day. Using the information in the table, design a menu/meal using only plant foods that would provide at least 12 milligrams of zinc.

## e-Vegetarians

www.sanitarium.com.au

**1** Vegetarians have a lower risk of dying from coronary heart disease. Explain using statistics.
**2** What are the reasons for vegetarians having a decreased risk of developing coronary heart disease?
**3** List the types of food consumed by vegetarians that are considered to have protective factors against the development of coronary heart disease.
**4** Why are they considered to have a protective effect?
**5** In the Western world, the cancer rate for vegetarians is lower than that for non-vegetarians. How much lower is the cancer death rate for vegetarians compared with that for non-vegetarians?
**6** What types of cancers are reduced by consuming a vegetarian diet?
**7** What food has a cancer-protective effect?
**8** List the major risks of high blood pressure.
**9** Why may vegetarians have a lowered risk of developing high blood pressure?
**10** Define obesity.
**11** Why are vegetarians less likely to become obese?
**12** Which food may decrease the chances of non-insulin-dependant diabetes?
**13** How does having a lower body mass index contribute to a decreased chance of developing diabetes?
**14** What is osteoporosis?

PULSES RISING

15 Which nutrient is implicated in increasing the chance of osteoporosis and hip fractures for non-vegetarians?
16 Why is this nutrient implicated?
17 List the other health risks of osteoporosis.

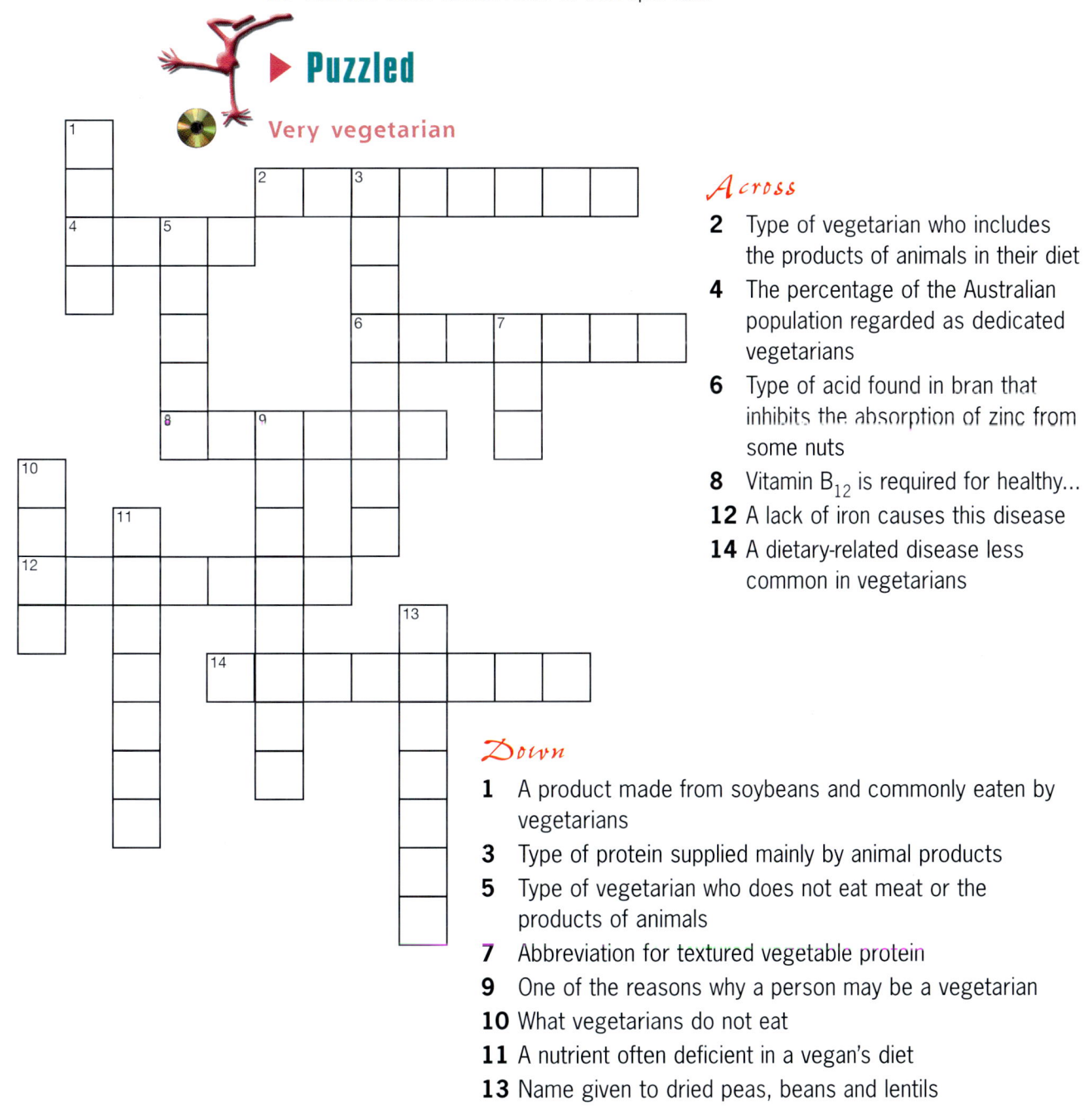

## ▶ Puzzled

### Very vegetarian

### Across

2 Type of vegetarian who includes the products of animals in their diet
4 The percentage of the Australian population regarded as dedicated vegetarians
6 Type of acid found in bran that inhibits the absorption of zinc from some nuts
8 Vitamin $B_{12}$ is required for healthy...
12 A lack of iron causes this disease
14 A dietary-related disease less common in vegetarians

### Down

1 A product made from soybeans and commonly eaten by vegetarians
3 Type of protein supplied mainly by animal products
5 Type of vegetarian who does not eat meat or the products of animals
7 Abbreviation for textured vegetable protein
9 One of the reasons why a person may be a vegetarian
10 What vegetarians do not eat
11 A nutrient often deficient in a vegan's diet
13 Name given to dried peas, beans and lentils

### True or false

1 Vegetarians do not eat meat.
2 A lacto-ovo-vegetarian eats fish, eggs, grains, milk and legumes.
3 A vegan eats grains, legumes, eggs, fruit and vegetables.
4 Most of the population in Australia are true vegetarians.

MEAT, FISH, POULTRY, EGGS, NUTS AND LEGUMES

5   Studies show that vegetarians tend to have a lower risk of developing obesity, diabetes and coronary heart disease.
6   Proteins from vegetable sources are classified as complete proteins.
7   A vegetarian diet is high in all vitamins, except vitamin B12.
8   A vegan tends to lack calcium in their diet.
9   Tofu is made from soybeans and is considered beneficial to your health.
10  Iron from vegetable sources is not absorbed readily into the bloodstream.

## Unscramble vegetarians

Unscramble the following letters to find out the names of famous vegetarians.

1  mot ceusir
2  herc
3  nolte honj
4  hotnany nibbsor
5  esscroat

# Meat, Fish, Poultry, Eggs, Nuts and Legumes: Assessment Task

This assessment task addresses the outcomes HPIP0501, HPIP0502 and TEMA0502 from the Health and Physical Education and Technology Key Learning Areas. The library research can be used as a basis to complete it.

▶ www.redmeat-feelgood.com.au

### Part 1 Meat Standards Australia

1  What is Meat Standards Australia (MSA)?
2  Identify and describe the three grades of meat.
3  What is the advantage of having the MSA label?

### Part 2 Increase in lean meats

In at least 250 words, discuss reasons why you think lean red meat has increased in popularity in recent years.

### Part 3 Poster

Teenage girls in particular are at risk of developing anaemia owing to a lack of consumption of red meat. Design a poster targeted at teenage girls to encourage their consumption of red meat, outlining the benefits of red meat and the problems associated with not including red meat in their diet.

### Part 4 Meat cuts

Select one meat cut and investigate a variety of recipes that use it. Select and design one of the recipes to suit your personal taste. Justify your design ideas for your new recipe.

## Section 7

# Water Water Water!

# Chapter 20

# Water Works!: Discover Water

**e-fact**
The average adult male consists of 40 litres of water.

**e-fact**
The human brain is 75 per cent water.

Did you know there is one thing we need even more than food each day? We need water! The average person can survive for at least a month without food, but most people will die if they do not have water for three days. About 60 per cent of the human body is made up of water.

We need to drink water so that all parts of the body can work properly:
- Water is essential for all of the chemical reactions in our bodies. For example, we need it for digestion to occur.
- It assists with the excretion of waste products from our body; therefore it is especially important for the functioning of our kidneys. These waste products are filtered by the kidneys and eliminated in our urine. Did you know that a lack of water is one of the reasons why people develop kidney stones?
- It lubricates our joints and protects our organs and tissues.
- It helps to regulate our body temperature. When we perspire, water evaporates from our skin, cooling us.

# WATER WORKS!

**@-fact**
Did you know that an elephant's trunk contains about 5 litres of water?

- It is an important part of many body fluids. For example, water is found in blood, which transports nutrients, oxygen and carbon dioxide around the body.

We should drink about 2 litres of water daily—about eight glasses of water—because we lose about 2–3 litres of water each day through breathing, perspiring and urinating. However, when it is hot or when we have been exercising, we need to drink more as we can lose up to 5 litres of water daily.

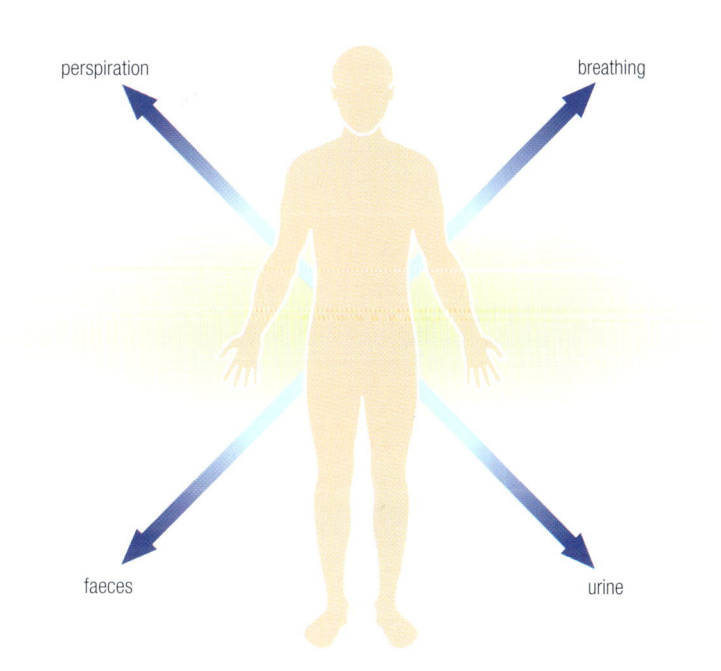

perspiration — breathing — faeces — urine

We should always drink water before, during and after exercising. A good rule is to drink a glass of water before exercising and half a glass every fifteen minutes. This helps to prevent dehydration and improves our performance.

Did you know that we obtain water through eating food? For example, milk contains 90 per cent water; watermelon, 93 per cent; celery, 95 per cent; and lettuce, 96 per cent.

Our fluid intake should be mostly made up of water, but we could include some milk, fruit and vegetables juices. Tea, coffee and alcohol should only be drunk in moderation because they tend to dehydrate the body.

We become dehydrated when we lose 1 per cent of our body weight in water. It is important to remember that when we become thirsty, we are already dehydrated. We should therefore endeavour to drink water throughout the day, rather than when we start to feel thirsty. Carrying a water bottle helps to prevent us from becoming dehydrated. Checking our urine is a good way to find out if we are consuming enough water: the more urine there is and the clearer in colour, the more hydrated we are. Being hydrated is being healthy.

**@-fact**
Dehydration means that the body is lacking water.

## Classify Water

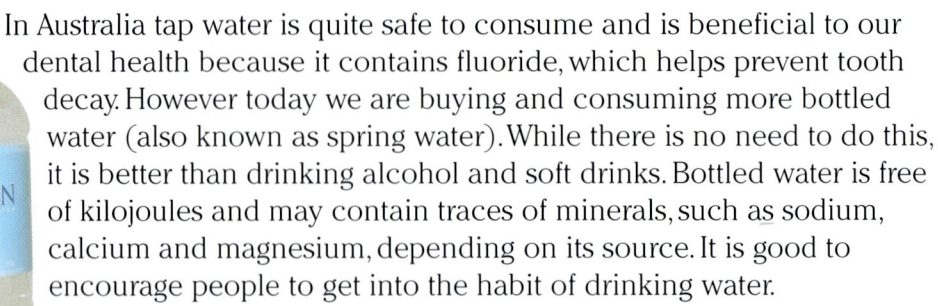

In Australia tap water is quite safe to consume and is beneficial to our dental health because it contains fluoride, which helps prevent tooth decay. However today we are buying and consuming more bottled water (also known as spring water). While there is no need to do this, it is better than drinking alcohol and soft drinks. Bottled water is free of kilojoules and may contain traces of minerals, such as sodium, calcium and magnesium, depending on its source. It is good to encourage people to get into the habit of drinking water.

Mineral water contains small quantities of mineral salts, like sodium chloride, sodium bicarbonate and magnesium. Unlike tap or bottled water, it is carbonated—contains bubbles. Years ago, mineral water was considered to have health benefits, but this is questionable because of its low mineral content. Nowadays, mineral water is bottled and drunk socially; it is often flavoured and sweetened and therefore similar in nutritional value to soft drink.

Soda water contains sodium, the amount of which depends on the brand. It was originally drunk for its health benefits, but today it is generally consumed as a mixer with alcohol, fruit juices or cordials. Soda water contributes to the overall sodium content of a diet and should not be consumed by those who need to restrict their salt intake.

## Focus on Water as a Nutrient

Water is considered to be a nutrient. It is kilojoule-free. It contains very small amounts of minerals, such as calcium, magnesium and sodium, and is therefore considered to be a nutrient.

## Water Products

### Carbonated water

The ancient Greeks and Romans bathed in natural mineral springs. The Europeans drank the naturally carbonated waters for their health. Carbonated water began being manufactured in 1772 in England, when British chemist Joseph Priestley discovered how to dissolve carbon dioxide in water. The carbon dioxide produces the typical sparkling quality and the bubbling effect known as effervescence. The use and popularity of this sparkling water had spread to America by the 1830s and later to Australia.

Sweeteners, fruit juice and flavours are now added to carbonated water. Such drinks are now enjoyed for refreshment rather than for their medicinal value.

Today, manufacturers create carbonated water by treating everyday drinking water to improve its appearance and to remove aroma and taste impurities and certain minerals. A device known as a carbonator is then used to dissolve the carbon dioxide in the treated water. Manufacturers may also add beverage syrup to make flavoured mineral water.

## Case Study: Neverfail

Harry Hilliam and his family established Neverfail in 1987. This company is said to be one of the pioneers of the bulk bottled water and cooler rental industry in Australia.

Neverfail uses good-quality natural spring water from underground sources. This water is naturally filtered through the Earth's rock formations. Stainless-steel tankers take it to the Neverfail factories, where it is generally processed and bottled within twenty-four hours. Unwanted contaminants, including bacteria, are removed through a filtration process at the factory. The water is then sterilised.

The plastic bottles are passed through a hot wash process for at least one minute to ensure that the inside and outside of the bottles are adequately clean. The bottles are then completely sterilised and rinsed to remove all detergents and sanitisers. Next they are filled automatically—without any human contact—and caps are put on within seconds.

The spring water is constantly checked and monitored throughout the entire process. It is held in-house for about forty-eight hours to ensure it is not delivered before the testing results confirm that it is free from contaminants.

In recent years, bulk bottled water sales have taken off. Neverfail has over 200 vehicles on the road, delivering water to its 150 000 customers.

### Questions

1. Where does spring water come from?
2. How is spring water filtered?
3. Describe how the plastic bottles are washed and sanitised.
4. Why are the bottles of water held at the factory at least forty-eight hours before distribution?
5. Draw a flow chart to illustrate the method of processing natural spring water.
6. How many times have you consumed bottled water in the past week?
7. Why do you think bottled water has become popular?

## ▶ Let's remember

1. Why do we need to consume water in our diet?
2. How is water related to the kidneys?
3. What percentage of the body is made up of water?
4. How does perspiring make us cooler? Explain by using diagrams and words.
5. How much water should we consume daily?

6 List the ways we lose water from our body.
7 Why do we need to drink more water when we have been exercising?
8 List three foods that are high in water.
9 Is bottled water better than tap water? Explain.
10 What nutrients are present in water?

## ▶ Let's investigate

1 **Investigate** the range of bottled water available in the supermarket. Categorise them as spring, mineral or soda. Why do you think that the sales of bottled water are increasing?
2 Look at the labels of spring, mineral and soda water. Tabulate the different nutrients present in each.
3 **Investigate** why water is referred to as $H_2O$.
4 Record your intake of water for a day.
   a Write down how many glasses of water you get from these sources:
      i tap water
      ii bottled water
      iii mineral water (state whether it is plain or flavoured)
      iv soda water
      v soft drink
      vi tea/coffee
      vii also list the high water-content foods you consumed
   b **Evaluate** your water intake.
      i Was your water intake adequate for your lifestyle and exercise program?
      ii How could your water intake be improved?
   c **Design** a plan to improve or maintain your water intake. For example, fill a water bottle each night and refrigerate. Place a reminder note on the fridge to take it to school.

## ▶ e-Water

www.softdrink.org.au

Go to the 'Health' section and click on 'Keep Your Cool'. Then visit the links 'Hot Days', 'On the Road', 'At Work' and 'At Sport'.
1 What causes the 'cool feeling' you experience?
2 What are the effects of dehydration?
3 Why does 'driving with liquids make road sense'?
4 How much liquid should you aim to consume every hour when travelling?
5 What helpful hints can you provide to help people consume liquids when travelling?

# WATER WORKS!

## ▶ Puzzled

### Who am I?

Unscramble the following letters to complete the sentence below: PEPCHEWS. Write the completed sentence in your workbook.

**e-HINT**
It is the name of a well-known brand of soft drink.

Jacob _____ was the first person to manufacture mineral water in England in the late eighteenth century.

### Number match-up

Match each statement in the first column of the table with the correct number in the second coloumn.

| Statement | Number |
| --- | --- |
| How many litres of water should we consume daily? | 2 |
| How many glasses of water is it recommended that you should consume daily? | 2–3 |
| What percentage of the body consists of water? | 3 |
| What percentage of the brain consists of water? | 5 |
| When exercising, how many minutes should elapse before you consume half a glass of water? | 8 |
| How many litres of water does the average person lose each day? | 15 |
| How many litres of water does a person who exercises lose each day? | 60 |
| How many days can you usually survive without water? | 75 |

### Mineral scramble

Unscramble the following minerals that are found in small amounts in water.
1. modusi
2. sangmimeu
3. matsopusi
4. droulefi
5. nori

### Aquatic poetry

Create an acrostic poem by using the term *water*. An example is provided below.

**W**onderfully refreshing
**A**quatic
**T**erribly cool
**E**xtremely essential for life
**R**eally healthy

WATER

# Water: Assessment Task

This assessment task addresses the outcomes HPIP0501, HPIP0502 and TEMA0501 from the Health and Physical Education and Technology Key Learning Areas. The library research can be used as a basis to complete it.

▶ **www.bottledwater.org.au**

1  The Australian Bottled Water Institute (ABWI) was established in 1995. Who is the ABWI?
2  What are the objectives of the ABWI?
3  What does consumer research suggest about the consumption of bottled water?
4  Discuss the following reasons why Australians are drinking bottled water:
   a  health conscious
   b  convenience
   c  taste
5  Collect a bottle of water and outline the information provided on its label.
6  Advertisers use the reasons listed in question 4 to promote the consumption of bottled water, despite the fact that tap water is safe to drink in Australia. Design your own advertisement for the launch of a new bottled water called AquaTaste.
7  Who sets the bottled water standards?
8  What regulations govern the bottled water industry?
9  Why is water considered important for one's health?
10 Draw a diagram to illustrate how water is gained and lost from the body. Indicate how much is lost and gained.
11 List the benefits of good hydration.
12 Research ways empty plastic water bottles can be recycled.

## Section 8

# Eat in Small Amounts

*The Australian Guide to Healthy Eating* sample serve is that which provides 600 kilojoules:

1 (40 grams) doughnut
4 (35 grams) plain sweet biscuits
1 slice (40 grams) plain cake
½ small bar (25 grams) chocolate
2 tablespoons (40 grams) cream, mayonnaise
1 tablespoon (20 grams) butter, margarine, oil
1 can (375 millilitres) soft drink
1 small packet (30 grams) potato crisps
⅓ (60 grams) meat pie or pastie
12 (60 grams) hot chips
1½ scoops (50 gram scoop) ice cream

# Chapter 21

# Fat Is Not a Four-Letter Word: Discover Fats

**e-fact**
Fats are also known as lipids.

The term *fats* refers to both fats and oils. At room temperature, fats are solid and oils are liquid. However we tend to just use the term *fats* when referring to either fats or oils.

While we are encouraged not to eat too much fat, our bodies do need fats in small amounts. They:
- provide us with the fat-soluble vitamins A, D, E and K
- act as a concentrated source of energy
- provide protection or insulation for body organs
- provide us with essential fatty acids

Fats have a variety of uses in food preparation, in terms of:
- Taste: They add flavour to food, for example, salad dressing.
- Moisture: They prevent food from drying out quickly, such as with bread or cakes.
- Texture: They provide a soft, crumbly or flaky texture, like scones.
- Appearance: They provide colour; for example, chips cooked in oil are golden brown.

FAT IS NOT A FOUR-LETTER WORD

**e-RIDDLE**

Q: What did the mayonnaise say to the fridge?
A: Close the door. I'm dressing.

- Cooking: They transfer heat evenly and quickly, such as with fried fish.

In recent years, it was believed that all types of fats were 'bad' and that we should reduce our fat intake. Now it is believed that some fats are beneficial to counteract conditions such as heart disease and arthritis. We refer to these as 'good fat'. The use of the terms *good fat* and *bad fat* are relatively new; monounsaturated and polyunsaturated fats are considered to be good, whereas saturated fats are bad.

It is important to remember that fats provide approximately twice as much energy as carbohydrates and protein. So to maintain a healthy weight, we should monitor our consumption of fats and choose monounsaturated and polyunsaturated fats in preference to saturated fats.

# Classify Fats

There are three main types of fats: saturated, monounsaturated and polyunsaturated fats.

## Saturated fats

Saturated fats are present in both animal products and coconut and palm oil. They should be avoided because they raise the level of cholesterol in the blood and contribute to heart disease. Food high in saturated fat includes butter, cooking margarine, lard (or dripping), meat fat, poultry skin, cheese, ice cream, commercial biscuits, cakes and pastries and takeaway foods.

To reduce our intake of saturated fat, we should:
- choose margarine instead of butter
- use vegetable oils instead of cooking margarine and lard
- eat lean meat and remove the visible fat
- avoid takeaway foods, which are deep-fried and often use palm oil or animal fat

## Monounsaturated fats

Monounsaturated fats should be eaten in preference to saturated fats. That way they can reduce both the amount of cholesterol in the blood and the risk of heart disease. For this reason we refer to them as good fat. Monounsaturated fats are present in canola oil, olive oil, avocadoes and nuts, such as peanuts, almonds and macadamias.

EAT IN SMALL AMOUNTS

## Polyunsaturated fats

Polyunsaturated fats are mainly present in plant foods and include sunflower oil, safflower oil, polyunsaturated margarine and nuts, such as walnuts, hazelnuts and Brazil nuts. Polyunsaturated fats include omega-3 and omega-6 fatty acids, which are considered to be essential fatty acids because they are necessary for the proper functioning of the body. They are also considered to be good fat. They are found in oily fish and polyunsaturated vegetable oils.

# Focus on Low-Fat Cooking

We have just learnt about the importance of choosing monounsaturated or polyunsaturated fats rather than saturated fats. Some additional tips to consider when preparing or cooking foods with oil are:
- Use as little oil as possible when cooking, so choose such methods as grilling, stir-frying, microwaving or steaming instead of shallow or deep-frying.
- Use a non-stick pan because it requires little or no greasing.
- Remove the skin or visible fat from meat.
- Be mindful of foods with invisible fat, such as cakes, chocolate and coconut milk.
- Read labels and look out for no-fat or low-fat products, or those that do not contain saturated fats.

### WEBExtras

**www.flora.net.au**
The Flora website provides lots of information about, among other things, the different types of fat, their components and the ways we can gain healthy lifestyles.

**www.foodsciencebureau.com.au/nutrit/quiz4/quiz.htm**
This Food Science Bureau webpage contains a quiz entitled 'Overweight and obesity'.

**www.devondale.com**
The Devondale website has loads of delicious recipes, nutrition and product information, as well as a kids' area that features fun, interactive games.

## ▶ Let's remember

1. What is the difference between fats and oils?
2. Why do our bodies need fat?
3. List the fat-soluble vitamins.
4. What is a lipid?
5. What are the uses of fats in food preparation?
6. What do the terms *good fat* and *bad fat* mean?
7. What kind of fat is thought to increase the risk of heart disease?
8. How can we reduce our saturated fat intake?
9. Why are monounsaturated fats considered good fats?
10. What types of food contain polyunsaturated fats?

# Chapter 22

# SWEET!: DISCOVER SUGAR

**e-fact**
Photosynthesis is the process whereby plants make carbohydrates from water and carbon dioxide.

**e-fact**
Did you know that biblical writings refer to the 'land of milk and honey'?

Would you believe that sugar was once called 'white gold'? This is because sugar cane did not grow that easily in the Mediterranean region. When fifteenth-century Italian explorer Christopher Columbus planted some sugar cane that thrived in the warm, rainy conditions of the West Indies, the Europeans battled for many centuries to get hold of the sugar-growing areas. Today, sugar cane is also grown in Australia.

Sugar is made in the leaves of the sugar-cane plant by a process called photosynthesis and is stored as a sweet liquid, or juice, in its stems. Cutting the stems and squashing them through heavy rollers separates the juice, which is then boiled to produce a thick syrup. From this syrup, raw sugar crystals are formed (sugar), leaving behind a dark, sticky liquid called molasses. Golden syrup is produced by partially breaking down the sugar into a liquid product.

EAT IN SMALL AMOUNTS

### e-fact
It is said that one of the admirals of Alexander the Great, king of Macedonia in the fourth century BC, described sugar as 'reeds which make honey without bees'.

## Classify Sugar

### e-fact
There is no advantage in consuming glucose drinks to get instant energy. The glucose passes from the stomach into first the small intestine and finally the bloodstream—a process taking several hours. It is not instant!

If you think of your favourite foods, it is most likely that some of them will contain some form of sugar. It could be the sugars found in either fresh fruits—such as pineapple, oranges or apricots—or cakes and breads. There are different types of sugars, depending on which food they are naturally found in. The most well known is sucrose, which comes from sugar cane. Glucose is the simplest form of all sugars and can be found in fruit and honey. It is also created when complex carbohydrates in cereal grains are digested. Other sugars include:
- fructose, which comes from fruit
- lactose, which comes from milk

### e-fact
Studies show that we are born with an innate desire for sugar.

SWEET!

### e-fact
Eating sugar activates a mechanism in the brain that releases the chemical serotonin, which makes us feel good.

### e-fact
Honey is one of the oldest foods. There is a Stone Age painting in the Tito Bustillo cave in Spain that portrays the earliest known record of honey being collected.

### e-RIDDLE
Q: What do bees do if they want to use public transport?
A: Wait at a buzz stop.

Although we should eat sugar in small amounts, it does play an important role in our diet by:
- adding taste, texture and colour to baked goods, such as cakes and biscuits
- providing energy for yeast in food, such as bread
- increasing the boiling point of food, such as confectionery, or reducing its freezing point, such as ice cream

## Sugar Products

There are many different sugar products. Some are listed in the table below.

| Product | Description | Uses in food preparation |
| --- | --- | --- |
| White granulated | Medium-sized white crystals, sometimes referred to as A1 sugar | Confectionery, preserving, beverages |
| Caster | Very small white crystals that dissolve quickly | Baking |
| Icing | Finely powdered white sugar | Icing, baking |
| Brown | Semi-moist, fine refined sugar | Fruit cakes, Christmas puddings, gingerbreads |
| Raw | Some processing has occurred but not fully refined | Beverages, some baking |
| Golden syrup | Concentrated, refined, golden-coloured sugar syrup | Baking |

white sugar    caster sugar    icing sugar

raw sugar    brown sugar    golden syrup

### e-RIDDLE
Q: What does a queen bee do when she burps?
A: Issues a royal pardon.

There is little difference in the nutritional content of the various sugar varieties. Sugar is a simple carbohydrate that is readily absorbed into the bloodstream to provide energy.

Sugar is present in fresh food and is added as an ingredient to many foods. It is better to obtain sugar in such food as apples and milk because you consume other nutrients besides simple carbohydrates. Foods such as

EAT IN SMALL AMOUNTS

## e-fact
Empty kilojoule foods are those that contain a lot of kilojoules (energy) and few other nutrients.

## e-fact
Brown sugar is not really a healthier alternative to white sugar. The only difference between the various sugars is the amount of molasses that remains on the crystal. So brown sugar is not more 'natural' than white sugar; it merely has a different colour and flavour and maybe contains traces of vitamins and minerals. The carbohydrate (and kilojoule) content is similar to white sugar. You would not eat brown sugar for its vitamin and mineral content!

lollies, soft drink and chocolates are referred to as empty kilojoule foods because they contain a high percentage of sugar and a low amount of other nutrients. We should therefore eat them in small amounts. The average annual sugar consumption in Australia is more than 50 kilograms per person. Over 80 per cent of this sugar is found in processed food and drinks! Can you believe that?

It is recommended to eat sugar in small amounts because:
- it contributes to dental caries
- it is not filling
- it makes food high in fat, such as ice cream and biscuits, more enjoyable and so encourages people to eat them and become overweight
- high-sugar food may displace other more nutrient-dense food that contains vitamins, minerals and fibre from the diet

sultanas          apricots          milk

## e-fact
It is a legal requirement that nutritional panels on food products contain information about 'Total Carbohydrates' and 'Carbohydrates in the Form of Sugars'. Sugars from milk, fresh and dried fruits and sugar cane must be included.

The nutritional panel on food labels lists the total amount of sugars. Other names for sugar on an ingredient list may include glucose, lactose and corn syrup.

The Australian Food Standards Code requires that when a food is labelled 'without added sugar' or 'no sugar added', the product must not contain any added sugars, honey, malt or its extract or maltose. It also sets down that processing must not increase the amount of sugars in the food. However this food may contain naturally occurring sugar.

## Focus on Artificial Sweeteners

## e-fact
Did you know that one can of soft drink contains seven teaspoons of sugar?

Scientists have invented artificial sweeteners to make food taste sweet without adding the kilojoules; they are used in a variety of food and drinks. However it is nutritionally more beneficial to increase the sweetness by adding food that naturally contains sugars, such as fruits and milk.

Saccharin was discovered in the late nineteenth century and is 300 times as sweet as sucrose. Most people will have a bitter aftertaste after consuming it. Aspartame is another artificial sweetener that was discovered in 1965. It cannot be used in baked goods because the sweetness is destroyed when heated. It is sold commercially as NutraSweet. Splenda—the brand name for a product derived from sugar—is chemically altered so that it does not contribute any kilojoules. It is 600 times as sweet as sugar and can be used in baked goods.

## ▶ Let's remember

1. Draw a flow chart to illustrate how sugar is produced.
2. What type of sugar would you use to:
    a. produce a sponge cake
    b. ice a sponge cake
    c. produce gingerbread
    d. produce toffee
    e. sweeten coffee
3. Is brown sugar better for you than white sugar? Explain.
4. Identify the different types of sugar found in:
    a. milk
    b. apricots
    c. malt
    d. sugar cane
5. Explain what an empty kilojoule food is and provide three examples of such food.
6. List some other terms that could be used on a food label instead of sugar.
7. What does it mean when a food label says 'no added sugar'?
8. What are artificial sweeteners and why do people consume them?
9. In your own words, explain what this statement means: 'High-sugar food may displace other more nutrient-dense food that contains vitamins, minerals and fibre from the diet'.
10. What is the advantage of using Splenda in baked products?

### e-RIDDLE

Q: How do you spell candy using just two letters?
A: C and Y.

## ▶ Let's investigate

1. The ending -ose usually indicates a type of sugar, such as in the words *glucose* and *sucrose*. Research the other types of sugars. Do they end in -ose?
2. Investigate the various types of artificial sweeteners on the market. Do you think these products promote healthy eating?
3. Read the labels of various 'diet' products available and compile a list of the various types of artificial sweeteners.
4. Study the ingredient list of a range of products that contain added sugar, and compile a list of the names of these added sugars.

EAT IN SMALL AMOUNTS

## ▶ Puzzled

### Sweet singing

Write down as many songs, or singers, that contain any of the words *sugar*, *honey* or *sweet*.

### Wonder sugar word

Find each of the words from the box in the puzzle.

aspartame
carbohydrate
caster
confectionery
dental caries
empty kilojoules
energy
fructose
glucose
golden syrup
honey
icing
lactose
photosynthesis
raw
saccharin
sucrose
sugar
sweet
white gold

| E | A | R | S | B | E | S | P | K | G | W | X | S | T | E |
|---|---|---|---|---|---|---|---|---|---|---|---|---|---|---|
| T | S | Y | H | A | U | S | A | E | H | N | E | F | M | S |
| A | P | W | R | G | C | D | O | I | N | I | A | P | G | O |
| R | A | A | A | E | C | C | T | R | R | E | T | N | E | T |
| D | R | R | B | Z | N | E | H | A | C | Y | R | S | S | C |
| Y | T | H | D | X | G | O | C | A | K | U | O | G | T | A |
| H | A | G | I | O | F | L | I | I | R | C | S | H | Y | L |
| O | M | M | L | I | A | N | L | T | U | I | L | O | L | R |
| B | E | D | V | T | S | O | H | L | C | H | N | N | V | N |
| R | T | S | N | M | J | K | G | B | G | E | B | E | V | I |
| A | M | E | S | O | T | C | U | R | F | N | F | Y | J | M |
| C | D | Y | U | C | A | S | T | E | R | E | I | N | Z | W |
| G | O | L | D | E | N | S | Y | R | U | P | G | C | O | F |
| T | E | E | W | S | L | K | Z | K | N | C | T | Y | I | C |
| S | I | S | E | H | T | N | Y | S | O | T | O | H | P | Q |

### Sweet sentence

Fill in the boxes below with the letters to make a sentence that relates to products sold commercially as NutraSweet.

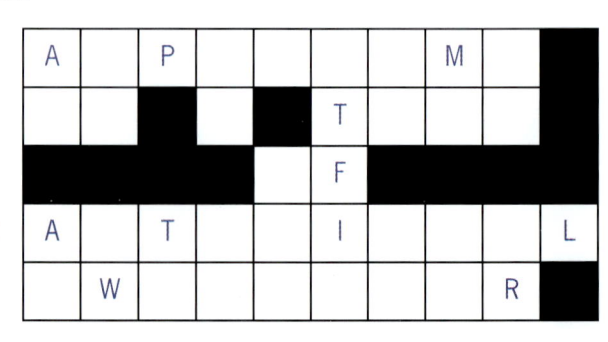

I W̶ I E T F A̶ E E
A S E A̶ P̶ T C X̶ E
S S T̶ I R E N M̶ A
A R P̶ A O T̶ Y P R̶ L̶

| A |   | P |   |   |   |   | M |   |   |
|---|---|---|---|---|---|---|---|---|---|
|   |   |   |   |   | T |   |   |   |   |
|   |   |   |   | F |   |   |   |   |   |
| A |   | T |   |   | I |   |   |   | L |
|   | W |   |   |   |   |   |   | R |   |

### Ukrainian proverb

*Only when you have eaten a lemon do you appreciate what sugar is.*

In your own words, explain what you think is meant by this proverb.

## e-Sugar

www2.csr.com.au

1. Click on the 'CSR Products' tab and **investigate** ten sugar products. Provide descriptions and uses for each. Ensure that you include a variety from each of the product ranges listed. You may wish to draw up a table like the one below to present your report.

| Product | Description | Uses |
|---------|-------------|------|
|         |             |      |
|         |             |      |

2. Click on the 'About Sugar' tab and then the 'Health & Nutrition' hyperlink. Write a 250-word report about how claims that sugar causes hyperactivity in children have not been proven, or select another topic to research and write a report.
3. Visit the 'Kids Room' tab and play the games to learn more about sugar.

### WEB EXTRAS

**www.bundysugar.com.au**
Bundaberg Sugar is a grower, miller, refiner and seller of sugar and related products in Australia. It owns and operates sugar mills in Queensland and is this State's largest cane grower.

**www.sugar.org.au/index.html**
The Australian Sugar Industry (ASI) Nutrition Information Service aims to provide scientifically accurate information on sugar nutrition and the role of sugar in the Australian diet.

**www.foodsciencebureau.com.au/nutrit/sugar.htm**
The Food Science Bureau website provides detailed facts about sugar, including the types of sugars found in food, an explanation of why sugar is added to food, how sugar fits into a healthy diet and the health implications of sugar.

# Chapter 23

# Salt of the Earth: Discover Salt

Salt is made up of the elements sodium and chloride. Sodium is found naturally in numerous foods, such as milk, eggs, meat and vegetables. We require this mineral for various functions, including:
- controlling blood pressure
- regulating muscle tone
- regulating the fluids that move in and out of cell walls

Salt = sodium + chloride

SALT OF THE EARTH

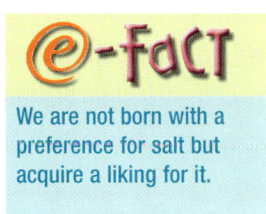

**e-fact**
We are not born with a preference for salt but acquire a liking for it.

In Australia we tend to eat too much salt in our diet as it is present in many processed foods. Did you know that Australians eat about 4000 milligrams of sodium daily? This is equivalent to about two teaspoons of salt per day. In actual fact, we should eat no more than 2300 milligrams of sodium, or one teaspoon, of salt per day.

A high intake of sodium can lead to many health problems, such as:
- dehydration
- high blood pressure (or hypertension)
- stroke

Salt is used as a preservative, additive and flavouring for food. Often, we add it to food during cooking and at the table; we also frequently add salt to food without even tasting it. People who consume a large amount often cannot taste the real flavour of food. They consider that low-salt or no-added-salt foods are too bland and tasteless. They have to be gradually weaned off salt so that they can appreciate the real food flavours.

*Alternative name for salt is sodium bicarbonate*

How can we reduce the amount of salt in our diet? Some suggestions include:
- Read food labels. Do you know other words for salt include sodium metabisulphate, monosodium glutamate, sodium nitrate, sodium bicarbonate, baking soda, vegetable salt and sea salt? Did you know that ingredients are listed in descending order of amount?
- Use lemon juice, herbs and spices as substitutes for salt.
- Taste the food before you add salt.
- Reduce the amount of salt used in cooking and added at the table.
- Choose products that say: low in sodium/salt, very low in

EAT IN SMALL AMOUNTS

sodium/salt, no sodium, low sodium/salt, reduced sodium/salt, lightly salted, salt-free, sodium-free and no salt.
- Reduce your intake of food that contains high levels of sodium, such as potato chips and savoury biscuits.

## Salt Products

### Sports drinks

Sports drinks are designed for people to replace their liquids and provide energy; some may contain minerals, such as potassium and sodium, to replace the minerals lost. Most people will not truly need to consume sports drinks unless they have exercised strenuously for at least one hour. After exercising, your body really requires fluids, not minerals or sugar. Water is actually the best drink for sportspeople.

▶ **Let's remember**

1. What minerals does salt consist of?
2. Why do our bodies require sodium?
3. List the dietary-related diseases that may result from consuming too much sodium?

**e-RIDDLE**

Q: What is a hedgehog's favourite food?
A: Pickled onions

4  Why do we add salt to our food?
5  Why do people who consume large amounts of salt have to be gradually weaned off adding salt?
6  a  List four types of food that contain salt naturally.
   b  List four types of food that have salt added to them.
   c  Discuss which foods are better sources of salt in our diet.
7  Identify some other terms to describe salt on a food label.
8  'Ingredients on a food label are listed in descending order of amount.' Explain what this statement means in your own words.
9  What terms should you look for on food packages to ensure it is low in salt?
10 Describe two ways you could help to reduce the amount of salt in your own diet.

## ▶ Let's investigate

1  Record your diet for one week. Include all meals, snacks and drinks.
   a  **Investigate** whether the food you consumed had added salt by reading labels and asking whether salt was added during the cooking process.
   b  List and total the food that had added salt that you ate for:
      i  breakfast  ii  lunch  iii  dinner  iv  snacks  v  drinks
   c  At which times did you consume the most food with added salt?
   d  At which times did you consume the least food with added salt?
   e  Were there any days, such as weekdays or weekends, when you tended to consume more food with added salt?
   f  Can you provide reasons why you ate food with added salt?
2  **Investigate** a range of food labels. Using your knowledge of alternative words to indicate the addition of salt, list the food products that have salt added to them.
3  As a class, **investigate** the taste of a range of food products that have a regular and a low-salt variety. Complete a table similar to the one below (some examples are listed). Indicate with an asterisk (*) the variety you prefer. Alternatively, the regular and salt-reduced products may be labelled 'Sample A' and 'Sample B' by the teacher so that you do not know which product is which and you have to work out the low-salt variety.

| Product | Taste of regular product | Taste of low-salt product |
|---|---|---|
| Tomato soup | | |
| Canned tomatoes | | |
| Tomato sauce | | |
| Soy sauce | | |
| Baked beans | | |
| Bread | | |
| Tuna | | |
| Potato chips | | |
| Peanuts | | |

EAT IN SMALL AMOUNTS

**4** Do you know which part of your tongue can taste salt? Refer to the diagram below. Place sweet, salty, bitter and sour food on various parts of your tongue.

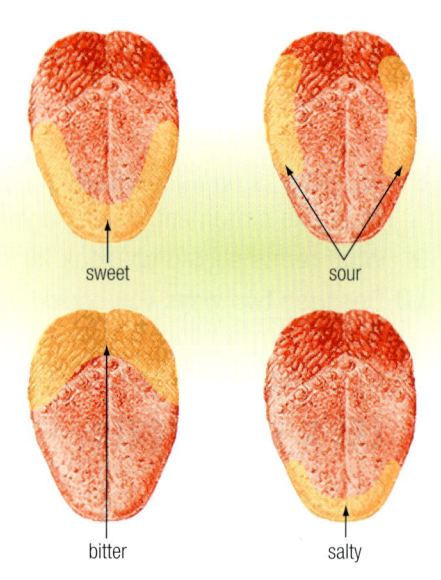

sweet    sour

bitter    salty

### WEBExtras

**www.gatorade.com.au**
Gatorade is a producer of a range of sports drinks and energy bars.

**www.gssiweb.com**
This is the website of the Gatorade Sports Science Institute, where endurance and sports science information can be obtained.

## ▶ Puzzled

### Salt scramble

Unscramble the following high-salt foods.
1. yos caeus (two words)
2. switesit
3. mdi smis (two words)
4. guessaas
5. neckchi noldeo supo (three words)
6. tatoop ipchs (two words)
7. ledsta nsut (two words)
8. gemtivee
9. nuta ni biner (three words)
10. idecklp sinono (two words)
11. coban
12. eamt eip (two words)

Can you think of a low-salt alternative for each of these foods?

### Potato chip investigation

Select two varieties of plain potato chips, one of which is low in salt. Your teacher will label them *Variety A* and *Variety B*.

#### Variety A

1. How well did you like the chips overall ☺ 😐 ☹?
2. How well did you like the appearance of the chips ☺ 😐 ☹?

3   The colour of the chips is:
    a   much too dark
    b   slightly too dark
    c   just about right
    d   slightly too pale
    e   much too pale
4   The texture of the chips is:
    a   much too hard
    b   slightly too hard
    c   just about right
    d   slightly too soft
    e   much too soft
5   How well do you like the flavour of the chips ☺ ☻ ☹?
6   The total flavour level is:
    a   much too strong
    b   slightly too strong
    c   just about right
    d   slightly too weak
    e   much too weak
7   I think the salt level is:
    a   extremely salty
    b   very salty
    c   moderately salty
    d   slightly salty
    e   not at all salty
8   How well did you like the aftertaste of the chips ☺ ☻ ☹?
9   How well did this sample meet your expectations for potato chips ☺ ☻ ☹?

### Variety B

1   How well did you like the chips overall ☺ ☻ ☹?
2   How well did you like the appearance of the chips ☺ ☻ ☹?
3   The colour of the chips is:
    a   much too dark
    b   slightly too dark
    c   just about right
    d   slightly too pale
    e   much too pale
4   The texture of the chips is:
    a   much too hard
    b   slightly too hard
    c   just about right
    d   slightly too soft
    e   much too soft
5   How well do you like the flavour of the chips ☺ ☻ ☹?

**6** The total flavour level is:
- **a** much too strong
- **b** slightly too strong
- **c** just about right
- **d** slightly too weak
- **e** much too weak

**7** I think the salt level is:
- **a** extremely salty
- **b** very salty
- **c** moderately salty
- **d** slightly salty
- **e** not at all salty

**8** How well did you like the aftertaste of the chips ☺ ☻ ☹?

**9** How well did this sample meet your expectations for potato chips ☺ ☻ ☹?

# Eat in Small Amounts: Assessment Task

This assessment task addresses the outcomes HPIP0501 and HPIP0502 from the Health and Physical Education Key Learning Area. The Heart Foundation of Australia website (www.heartfoundation.com.au) can be used as a basis to complete it.

1. What is the tick program and why was it established?
2. The tick program has nutritional guidelines for ten different categories of food. What are they?
3. 'Eat Smart for Heart' and 'Jump Rope for Heart' are two other programs developed by the Heart Foundation. Describe these programs and analyse how they may enhance the health of individuals and populations.
4. What is the difference between being overweight and obese?
5. What health problems can be linked to obesity?
6. How common is high blood cholesterol in Australia?
7. Look at the section on low-fat tips when buying takeaway food. Think about the choices you make when you buy such food. What are three tips you could follow next time?
8. Australian children are becoming more sedentary rather than participating in physical activity. Do you agree or disagree with this? Include the use of statistics in your answer.
9. Write down a list of food you consume regularly that you consider to be high in fat? Make three suggestions as to how you could modify your diet.
10. What resources are available to assist you with making healthier food choices?

# GLOSSARY

**amino acids** organic compounds that occur naturally in plant and animal tissue and that form the components of proteins

**antioxidant** a food additive that prevents the oxidation of fats

***Australian Guide to Healthy Eating*** a food guide that indicates suggested serving sizes from the Five Food Groups

**carbohydrate** any of a large group of energy-producing organic compounds that are made up of carbon, hydrogen and oxygen

**cure** to preserve food though various methods, such as salting, drying and smoking

**fibre** the part of food that cannot be digested and absorbed to produce energy

**folic acid** a vitamin of the B complex present in leafy green vegetables, liver and kidneys; also known as folate or folacin

**fortified food** food that has vitamins added

**Glycaemic Index** a measure of the rate at which carbohydrates are broken down into glucose

**Healthy Eating Pyramid** designed by Nutrition Australia and provides a guide as to the proportions of food we should eat, which we should eat in moderation and which we should eat in small amounts

**iron** an important element in the blood that forms part of the haemoglobin that gives red blood cells their colour

**isoflavone** a naturally occurring compound present in chickpeas and legumes, such as soy, which has the most concentrated amount

**lactose** the name of the sugar found in milk

**lactose intolerant** when a person's body cannot produce enough of the enzyme lactase, which breaks down lactose so that it can be absorbed by the body

**legumes** pod-bearing plants, such as peas and beans; also known as pulses

**lycopene** a natural red colour in food, such as tomatoes

**metabolism** the sum of all chemical reactions that keep the body functioning

**nutrient** a chemical compound found in food that is essential to health and life

**omega-3 fatty acids** unsaturated fatty acids that are present mainly in fish oils

**osteoporosis** a condition where the bones become brittle and can break

**pectin** a type of insoluble fibre that can be found in varying amounts in different fruit

**phytochemicals** the chemicals found in plants that are said to provide protection against diseases, such as cancer, osteoporosis and heart disease

**polyunsaturated fats** fats that have two or more double bonds in the linked carbon chain that are able to admit two hydrogen atoms; they are not associated with the accumulation of cholesterol

**protein** any of a large group of organic compounds made up of amino acids and found in all living things

**rancid** when a food develops an unpleasant and 'sour' taste

**RDI** stands for recommended daily intake

**UHT milk** milk that has been sterilised by heating to a very high temperature for a short period of time to kill all of the bacteria present

**vegan** a person who does not eat meat or animal products

**vegetarian** a person who does not eat meat but may eat animal products, such as eggs and milk

# INDEX

artificial sweeteners 204–5
*Australian Guide to Healthy Eating, The* 26–9, 35, 113

beans *see* legumes
bok choy 75–6
bread 41–2
   classification 43
   products 44
   properties 42

calcium 114–15
carbohydrates 66–7
cereals
   classification 33
   products 35
   properties 34
cheese 129
   classification 130
   products 131
   properties 130
chicken *see* poultry
cholesterol 168–9
cultural factors 13–15

diet and vegetables 76
Dietary Guidelines for Australians 25–6
drinks, sports 210

e-food models 28–9
economic factors 15
eggs 166–7
   classification 167–8
   products 169
   properties 168

fats 198–9
   classification 199–200
   low-fat cooking 200
   monounsaturated 199
   polyunsaturated 200
   saturated 199
fibre 34–5
fish 148–9
   classification 149
   products 150
   properties 149
folate 79–80
fruit 102–3
   classification 103
   products 104
   properties 104

grains 32–3
   *see also* cereals

Healthy Eating Pyramid 24–5

iron 143

legumes 89–90
   classification 91
   products 92–3
   properties 90–1
lycopene 78–9

meat 138–9
   classification 139–42
   products 144
   properties 142–3
   types of cuts 139–42
media influences 19–22
milk 112–13
   classification 113–14
   products 115
   properties 114

noodles 64
   classification 65–6
   products 67–8
   properties 66
   types 66
nutritional knowledge 16–18
Nutrition Australia 24
nuts 174–5
   properties 176

omega-3 150

# INDEX

pasta 64
　classification 65–6
　products 67–8
　properties 66
　types 65
phytochemicals 91
potatoes 77
poultry 156–9
　classification 160
　products 160–2
　properties 159
protein, incomplete 176

religious factors 13–15
riboflavin 130–1
rice 50–1
　classification 53–4
　cooking 52–3
　processing 51–2
　products 55
　properties 53

safety 2, 6–7
salt 208–10
　products 210
seafood *see* fish
social factors 12–13
soy 92
soy milk 93
sports drinks 210
sugar 201–2
　classification 202–3
　products 203–4

supplements, vegetarian 184
sweeteners, artificial 204–5

textured vegetable protein (TVP) 93
tofu 92

Vaalia 122
vegetables 74–5
　Asian 75–6
　classification 79
　and diet 78
　properties 78–9
　salad 76
vegetarian supplements 184
vegetarians 182–4
　classification 182–3
vitamin C 104

water 190–1
　carbonated 192–3
　classification 192
　as a nutrient 192
　products 192–3

yoghurt 120–2
　classification 122
　products 122
　properties 122

# RECIPE INDEX

American cheeseburger 146
American hotcakes 39
apple and rhubarb crumble 108

banana and strawberry shake 128
banana smoothie 108
beef Hokkien noodles 69
beef satay 147
berry whip 133
bok choy and lamb stir-fry 88
bruschetta 47
butter masala chicken 164

California rolls 63
calzone 48
cheese and bacon risotto 62
cheesy crumbled chicken strips 49
chicken and pear 109
chicken burger 165
chicken filo parcles 163
chicken laksa 70
chicken wraps and hummus 98
chilli chicken with nuts 181
chilli con carne 147
choc nut cookies 180
chocolate bread and butter pudding 49
chocolate cherry dessert 118
chocolate raspberry muffins 39
chocolate soufflé pudding 172
coconut and yoghurt fruit salad 128
creamy chicken curry 62
crispy baked fish 154

damper 47

easy cheesy cannelloni 70
easy individual pavlova 108

felafels 98
fettuccine carbonara 172
focaccia melt 134
French toast 171
fresh fruit and nut pastries 180
fruit and oat cookies 127
fruity yoghurt pancakes 127

gnocchi 87

hoisin drumsticks 164

Indian fish curry 154
Irish soda scones 118

jacket potato 133

lamb cutlets with yoghurt and couscous 128
lemon and poppy seed muffins 126

macaroni Waldorf salad 69
make your own yoghurt 126
Mexican nachos 40
minestrone soup 69

passionfruit ice cream 119
peach pillow 109
peach trifle 119
peanut butter 181
penne and chicken bake 118
pork and prunes 146
potato and ham frittata 173

rice and chicken wraps 61

savoury cheese muffins 134
spaghetti marinara 155
spicy lamb, vegetable and coconut pilaf 61
spinach and feta pie 171
spinach and leek quiche 134
sweet curried fish 153

tuna fish cakes 153

veal schnitzel 146
vegetable frittata 88
vegetable fritters 87
vegetable korma 99
vegetable lentil curry 99
vegetable risotto 88
vegetarian curry couscous 39
vegetarian triangles 40

warm chicken salad 165